EDIBLE WEED APPRECIATOR

6. BE MINDFUL OF POLLUTANTS These could include traffic fumes, stormwater run-off, polluted water bodies, industrial pollutants, and lead paint flaking from house walls.

7. BE MINDFUL OF HERBICIDES Check your local laws to be sure, but most councils require that sprayed areas are marked with signs. Some municipalities colour their herbicides with dye.

8. BE PERSISTENT Don't be put off by one experience with a weed that's been too tough, sour or bitter for your taste. It may taste quite different growing in a different spot.

9. EXPERIMENT with ways to use your newfound culinary ingredients.

10. RELISH the experience!

THE WEED FORAGER'S HANDBOOK

THE WEED FORAGER'S

HANDBOOK

A GUIDE TO EDIBLE AND MEDICINAL WEEDS IN AUSTRALIA

**ADAM GRUBB &
ANNIE RASER-ROWLAND**

HYLAND HOUSE PUBLISHING
MELBOURNE

First published in Australia in 2012 by Hyland House Publishing Pty Ltd
www.hylandhouse.com.au

Reprinted 2013, 2014, 2015, 2016, 2017, 2019, 2020, 2021, 2022, 2024

National Library of Australia Cataloguing-in-Publication entry:
Author: Grubb, Adam.
Title: The weed forager's handbook: a guide to edible and medicinal
                   weeds in Australia / Adam Grubb and Annie Raser-Rowland.
ISBN: 9781864471212 (pbk.)
Notes: Includes bibliographical references and index.
Subjects: Wild plants, Edible–Australia.
                   Medicinal plants–Australia
                   Weeds–Australia.
Other Authors/Contributors:
                   Raser-Rowland, Annie.
Dewey Number: 581.6320994

This 2024 reprint printed by CSIRO Publishing under a rights licence.
www.publish.csiro.au

Editing by M Schoo
Design and layout by Rob Cowpe Design
Printed in China by 1010 Printing International Ltd

The paper this book is printed on is in accordance with the standards of the Forest Stewardship Council® and other controlled material. The FSC® promotes environmentally responsible, socially beneficial and economically viable management of the world's forests.

MIX
Paper | Supporting
responsible forestry
FSC® C016973
www.fsc.org

FOREWORD

We will fight them in the fields, we will fight them in the gardens, we will fight them in the footpaths. We will fight them why? Because someone told us to. Because someone taught us to react to a word, to a definition. Our immediate conscious action when someone mentions the word weed is a reaction. And this reaction combines additional vocabulary with the word weed, such as invasive, thus building the siege mentality to the point that weeds have become a war. Weeds have become a win-at-all-costs battle that has stepped beyond the boundaries of the rules of war. There is no Geneva Convention when people enter the mindset of the world war on weeds. It's an anything goes barrage. There is no 'below the belt' when it comes to this rumble.

Weeds have been a part of my life. Growing up with a Greek heritage, gardening was a given at every residence I ever visited, be they aunts, uncles, great distant relatives, third cousins or just random Greeks for that matter. Growing was a God-given blessing and everyone grew productive and beneficial plants in whatever they could get soil into. And weeds were a day-to-day part of the plant vocabulary, the best part being that they were not seen as weeds. I always remember bus trips and picnics, and the coach or car would pull up on the side of the Hume, the Pacific Highway or by the local park and there would be all the old aunts and *yiayias* (grandmothers) walking along the side of the freeway 'emu-bob' style, pivoted at the hips and collecting every kind of so-called weed you could not imagine. When the bus stopped at the end of the trip, people lined up to pick up their bag of horta or weeds, call them what you like. These would then be turned into the most incredibly tasty and nutritious hortopita or weed pies (like spinach pie but a weed pie). Imagine the wild fresh greens, picked only an hour or two earlier being turned into that night's dinner for the family. It sings the song of reduced footprint, resilience and wild diversity in one broad brush example.

Every footpath, park, or highway was not seen as some sort of grass maintenance nightmare. No no. It was seen as one big free wild salad bowl. And this is a good way of looking at this book. It is a crazy big bowl of information on plants that you would otherwise be encouraged not to consider edible, or beneficial medicinally or nutritionally. I am so pleased that more and more people are seeking out the knowledge and truths of the village and taking guided weed walks through their local area. This information and experience of the ages is getting the recognition that it duly deserves.

The book looks at the very nuts and bolts of weeds. It is a wonderful combination of scientific description, seasonally applicable take-home information, clear photographs and drawings and equally important cultural story telling. It reflects on weeds as colonisers. They are the mongrel street fighters that come along and re-establish life where there is only death and desolation. The capacity of weeds to set up shop and then in essence create the conditions for the next round of settlers or succession is quite remarkable. They are the true pioneers of soil building, bringing life and tolerating the torment that our technology and development inflicts. They adapt to disturbance readily, and given that we humans are such a disturbance, then it stands to reason that we go together with weeds much like a bum goes together with a pair of underpants (*kolo kai vraki* as we say in Greek).

So cast aside those weed goggles and open up your eyes and your mind to the wonderful world of weeds. As famed organic farmer Joel Salatin would put it, allow that amaranth to express its amaranth-ness. Let all weeds express their specific and unique weed-ness. Allow observation and information to open up a whole new vocabulary of wonder in the world of weeds. Don't react, rather respond to the endless possibilities that plants (formerly known as weeds) can offer you. From weekend gardener, school gardener, community and verge builders, foragers and scrumpers, urban guerrilla planters or humble rooftop and balcony farmers, this book is a must for all growers of our future food and health security. In other words, if you eat, then this book is an essential companion.

Costa Georgiadis
BLArch, CompOSTA, Weed Ingester

Annie used to make art but became suspicious that the natural world was frequently outdoing her, and with greater finesse. She took up making gardens for beauty, then realised that she could feed herself at the same time. These days she is a graduate of horticulture and permaculture certificate courses, has worked at several Melbourne nurseries, done urban permaculture design and spent time in Tanzania working on permaculture systems. As a full-blown plant nerd with a passion for questioning modern food equations, she started investigating wild and other undervalued foods, was recruited by the rich pleasures of foraging and gleaning, and hasn't stopped learning since.

Adam left a career in IT to pursue becoming backwardly mobile after doing some research into energy depletion that lead to a small crisis of confidence in civilisation. In 2003 he founded the energy news clearinghouse Energy Bulletin which became the most popular website about peak oil on the net. The move to energy writer did nothing for his physique however, so he got into permaculture in 2004 and co-founded (with Dan Palmer) the now global permablitz movement. These days he and Dan are co-directors of the urban permaculture consultancy Very Edible Gardens (VEG) in Melbourne. He has been stalking through the shrubbery collecting wild foods for almost a decade.

CONTENTS

Acknowledgments

Thanks to Frances and Sarah for their feedback, and to the dog for putting up with fewer walks.

Photo and illustration credits

We have made extensive use of botanical illustrations from archival texts. The vast majority have been made available thanks to the impressive efforts of Kurt Stüber at the Max Planck Institute for Plant Breeding Research (www.biolib.de).

A notable exception is the prickly pear illustration by Mary Emily Eaton from *The Cactaceae* (1919), made copyright free thanks to the Carnegie Institute and www.cactuspro.com.

The angled onion illustration from *Flora Graeca* (1806) has been made available by the Sherardian Library of Plant Taxonomy, one of the Bodleian Libraries of the University of Oxford.

The wild brassica illustration from *Flora Danica* (1806) is courtesy of The Royal Library of Denmark.

The photo of Theseus and the Bull of Marathon on a terracotta pot from circa 440-430 BC, held in the Metropolitan Museum of Art, New York, was taken by Marie-Lan Nguyen, and made available on commons.wikimedia.org under the Creative Commons Attribution 2.5 licence.

The cover illustration is from *Herbarium Blackwellianum* by Elizabeth Blackwell (1757).

The photo of the authors foraging is by Louis Fourie.

All other photographs are by Adam Grubb.

1 | ON THE APPRECIATION OF WEEDS

There are laws in the village against weeds.
The law says a weed is wrong and shall be killed.
The weeds say life is a white and lovely thing.
Carl Sandburg (1922)

Weeds are the ultimate convenience food. They ask of you no money, no search for a parking space at the supermarket, no planting, no watering or any other maintenance whatsoever. Gathering them may call for a walk to the park before dinner. A walk that gets you a little more exercise that

Fleabane, doing some vertical gardening all by itself.

day, your face deliciously rain-wet or your shoulders gloriously sun-warmed and your appetite really roaring.

A remarkable number of the common weeds that sprout from Australian gardens, abandoned lots, parks and cracked footpaths, are either edible, medicinal, or both. Among them are some of human cultures' oldest and once most beloved food plants, and some of the most nutritious plants ever tested by modern science. We've documented our favourites here, in the hope that it will help you to identify, appreciate and use them safely, as free, vitamin packed, environmentally low impact fare.

Collecting wild foods is deeply rooted in our nature. For the vast majority of human history we have been hunter-gatherers. Even as most of our ancestors became agriculturalists over the last 10,000 years, wild foods remained an important part of their diet. Reconnecting with the original function of our foraging impulses helps us satiate them before they erupt into a house full of unused kitchen gadgets, shoes, or an LP jazz collection!

Foraging heightens the senses, yet is simultaneously relaxing – almost hypnotic. Urban foraging often leads you into semi-wild, overlooked places – secret lots and abandoned blocks where, within the metropolis, nature is reclaiming some ground.

But are weeds a rightful part of 'nature'? What is a weed anyway? Weed science textbooks offer indefinite definitions such as 'a plant that according to some human criteria, is undesirable'. In 1956 Professor William Stearn put it this way: 'Taken as a whole, weeds are not so much a botanical as a human psycho-logical category.'

Plants once considered of vital cultural, culinary or medicinal importance can fall from favour, and become despised 'weeds'. It's just as possible for them to traverse their way back into our sympathies. Hence, an equally valid way of doing the weeding,

A well utilised manhole cover

rather than with hoe or spray bottle, is to look past the tarnish of cultural derision, notice the virtues of these maligned plants, and perform a mental reclassification. For a 'valuable weed' is an oxymoron.

That's not to say we don't sometimes pull weeds from the garden – we do. But first we ask: is this plant suggesting something about the nutrient or moisture levels in this part of the garden? Is it protecting or building soil, or assisting in natural pest control? Can I eat it? Can it make my locks look lustrous?

Given that so many of the plants listed in Chapters 2 and 3 have long histories with humanity, myriad uses for them have been discovered. They tend to follow settled humans all around the world, and for a simple reason: most declared weeds are plants adapted to disturbance – including the kinds we humans are so adept at, such as clearing, ploughing and polluting. These plants have evolved to follow in our wake, and we may, by long association, have evolved the capacity to eat them. One of our staple salad greens, sow thistle, adorns the cover of the book *The Worst Weeds of the World* (Leroy Holm, 1977). It lists the world's 18 most objectionable weeds. A full 16 of these can be used for human consumption.

You will not get your full diet from weeds, since most are leafy greens, high on nutrients, low on energy. (So don't give up your day job just yet.) However, they can be the perfect antidote for the high energy, low nutrition foods that characterise the Western diet. In extreme times such as war, including the four-year siege of Sarajevo, weeds have helped avert nutritional catastrophes. Eaten fresh, weeds provide living enzymes and dense vitamins and minerals. Wild plants in general are richer than cultivated plants in vitamins A, C, E and K, antioxidants and omega-3s. Cultural use of wild foods has been associated with many health benefits, not the least of which is anti-ageing.

In producing these free, nutrient-rich foods, no inputs are required. No fossil fuels are burnt in refrigerated trucking. No water is drained from fragile river systems. The seas are not polluted with fertiliser run-off. No birds die eating crickets that try to feed on pesticide-coated leaves. It may not be possible to eat a more sustainable food.

While weeds may provide the ultimate sustainable dinner, given their reputation as environmental villains this virtue might seem one bright anomaly in an otherwise gloomy scene. Yet even their bleak environmental reputation is increasingly questioned. Weeds tend to live fast, procreate a lot, and die young, characteristics of 'pioneer species' – plants adapted to colonising disturbed earth. Weeds are often the only species vigorous enough to take hold in highly damaged ground, thereby stemming erosion, rebuilding soil fertility and structure, moderating climate, and creating habitat. Where disturbances are not repeated, this process of rebuilding can eventually lead back to 'climax' ecosystems such as old-growth forests or grasslands.

It turns out that humans, with our mining, construction, clearing, ploughing, herbicides, and water channelling, create a lot of highly damaged ground. Some ecologists now suggest that in many situations working with weedy species, rather than against them, is the only sustainable way to begin repairing some of the (very extensive) disturbances that we have left in our wake.

Notes of Caution

Poisonous Plants!

To the untrained eye, many poisonous plants are difficult to tell apart from edible and medicinal plants. Some plants are toxic even at very low doses. Below are a few to be wary of, but naturally there are countless others. The rule is not to eat any plant unless you are sure of its identity, and that it is not toxic.

Danger

CASTOR OIL PLANT (*Ricinus communis*) With an adult lethal dose being as little as four chewed seeds, this is a plant to be super wary of. The large, lobed leaves give it a tropical look (and make it hard to confuse with any of the weeds covered in this book). It is indeed widespread in tropical Australia, though also present throughout the rest of the country.

HEMLOCK (*Conium* species) Once used for killing ancient philosophers – well, Socrates anyway. To the unpractised eye, hemlock could possibly be confused with parsley, fennel or wild celery (to which it is related), so practise extra caution with plants in this family. It is very toxic. If you get sap on you, wash it off immediately.

ASTHMA WEED or PELLITORY (*Parietaria judaica*)   This is a very common weed, sometimes associated with allergic reactions. Yet weed-loving Sydney artist Diego Bonetto tells us that this is one of his favourite greens. He has eaten this plant (picked before it goes to seed) for years, apparently with no ill effect, and told us that it appears in old Italian recipes. However, many people do suffer hay fever from the pollen or get rashes from touching this species.

ALLERGIES, AND NEW FOODS

Any new food should be introduced into your diet gradually, to make sure it works well with your body, and to give your digestion time to build up the necessary enzymes. If you suffer from any existing allergies, be especially careful, particularly with plants closely related to those to which you have the allergies.

Nitrates

Several of the plants in this book appear on 'toxic plants for livestock' lists because of their ability to accumulate nitrates, especially when grown with artificial fertilisers or on un-composted manure. These include fat hen, mallow, dock, purslane and wild brassicas. Among cultivated vegetables, spinach, lettuce and broccoli are all potential nitrate accumulators too. Nitrates generally aren't a problem for adult people (you would have to eat unusually large amounts of greens before there was any toxicity) and they may well account for some of the health benefits of vegetables. However, they are more toxic to infants, so you should not feed puréed greens of nitrate-accumulating plants to children under six months.

Oxalic Acids

Many wild edibles have higher levels of oxalic acids than cultivated vegetables. The ones in this book which we know to contain high levels are amaranth, dock, fat hen, oxalis and purslane. Oxalic acids occur naturally in high levels in many common foods, including almonds, chocolate, bananas, rhubarb, parsley, beer, tea and spinach. Even if we don't eat them, our bodies create oxalic acids from other sources, and some scientists suggest that they play as yet little-understood roles in the body. However, when oxalic acid combines with calcium and some other minerals it creates crystals which in some people can contribute to kidney stones, gout and rheumatoid arthritis. Too much oxalic acid in the diet can limit our absorption of calcium. Doctors advise that pregnant women and anyone with a predisposition to any of the above conditions should limit their intake of oxalic acids. Foods high in both calcium and oxalic acid are less of a concern than those just high in oxalic acid, and most of the weeds mentioned (with the known exception of dock and possibly oxalis) also have high calcium content. We basically use three strategies for dealing with oxalic acid: blanch for five minutes and dispose of the water, or combine the plants with high calcium foods, as in our purslane tzatziki recipe. The third is to be sensible. Don't let your newfound fervour for weeds lead you to eat entire buckets of fat hen in a single sitting.

PREGNANCY

A special note to pregnant women and those trying to conceive: there are warnings against eating large quantities of many plants containing strong flavour compounds (including basil, parsley and cinnamon), as there is a possibility of stimulating uterine contractions. Weeds that fit this description include oxalis, horehound and shepherd's purse.

CONTAMINANTS

While the plants in this book may themselves be highly nutritious, you need to be observant and use common sense when harvesting them. There are three main types of contaminants to consider: chemical (primarily herbicides and pesticides), heavy metal, and biological.

Be alert for signs of herbicide wherever you are picking weeds. Some councils use a coloured dye and/or signs to indicate sprayed areas. Get to know your local area's spraying regimes if you are foraging in public spaces, and check with the council if uncertain.

Ironically, the places that get sprayed more regularly are often the places where you will find more edible weeds, as they return to fill the void that is naked ground. In areas less sprayed, the weeds are usually shaded out by longer-living species which are exploiting the soil-building effects of the weedy colonisers.

Perhaps more insidious than herbicides are industrial chemicals and heavy metals, and it is due to these that we choose not to pick weeds growing too close to urban waterways or former industrial sites. Of particular concern is lead, which has been added to our soils courtesy of lead paint, vehicle exhaust, and other household and industrial sources. For this reason we also don't forage near busy roadways or other places where the soil might be highly contaminated with lead. Anywhere near old houses where lead paint may have flaked off or been sanded is worth thinking twice about. Lead is especially concerning for young children as it can affect their development. The ability of many weeds to extract high levels of nutrients from the soil may correlate with higher levels of lead uptake. Despite this, all plants are actually pretty good at not taking up lead. Usually it is the dust on them which is of more concern, so it's important to wash leaves well when harvesting in suspect areas – a dash of vinegar in the water helps remove any external heavy metals. After years of almost daily weed eating, Annie recently had a blood test that showed completely negligible levels of lead.

The third type of contamination is biological. You probably got the hang of not eating faeces fairly early on in life, and while dog (and human) urine is essentially sterile, it may not be a great salad dressing. For all the above reasons we recommend washing your weeds, particularly in the city.

2 | OUR TOP 20 EDIBLE AND MEDICINAL WEEDS

In compiling these notes we considered for the first time just how many species of weedy plants we personally use for food or medicine, and were amazed when the count passed fifty. In this section we've listed our absolute favourites. Many are seasonal but, as seasons overlap, you can find most of them at any time of year. While most are available throughout the country (and indeed much of the settled world), we admit to an unavoidable bias towards the south-east, because that is where we live. All the main weeds listed can be found in Adelaide, Canberra, Hobart, Melbourne and Sydney, and all but one in Brisbane and (subject to rainfall) all but two in Perth.

IDENTIFICATION NOTES

While we've included our own pictures, it's also very useful to see each plant at different stages of life and from different angles. Absolute identification would require familiarity with taxonomic botany, but we've tried to note where look-alike species occur. We encourage you visit our website www.eatthatweed.com for links to online images of the weeds we have covered, and also to find details of our Edible Weeds Walks. If you are in any way uncertain, DO NOT EAT the plant, and be aware that even many edible plants have both edible and poisonous parts, including some mentioned in this book. Research anything you are unsure about, and be sure to read our Notes of Caution in Chapter 1 to familiarise yourself with a few other possible foraging hazards.

General tasting notes

The first time you taste a new food, your tongue is naturally suspicious, especially of bitter flavours. Those of you who remember your first tastes of beer or coffee will know what we mean. It's often only after you eat a new food, sleep, and wake up alive and well, that your tongue is willing to appreciate its nuances. So, if you are unsure of how you feel about the flavour at first (but are sure of your identification), do try again at a later date. As a general note, because weeds are not bred for the human palate, choose the youngest, lightest green, and freshest looking leaves, as these will be the least bitter or fibrous. Greens are also usually much tastier from plants yet to flower, in some cases dramatically so. If you have ever grown a lettuce and tried to harvest some leaves after flowering you will well know this transformation! With regards to texture, it often helps to avoid using stems, and to chop leaves finely, against the direction of the fibres.

Medicinal notes

Many of the medicinal plants mentioned here are gentle 'nutraceuticals', while others contain more powerful compounds to the point of being hazardous if used without respect. Using herbal medicines, with free ingredients, allows you to take some of your health problems into your own hands. That said, we've had mixed results ourselves with herbal remedies prescribed by alternative practitioners or found in herbal remedy books – perhaps because there can be a tendency to repeat the best and worst of folkloric traditions without question. For this reason, we have taken the same approach in writing this book as we do in our personal lives: alongside investigating traditional uses we've searched the scientific literature for studies about each plant. Since these plants can be harvested for free, pharmaceutical giants have little incentive to fund studies into them. Even given this, there are many independent studies and much good news to report, and what we have mentioned in these profiles likely represents only a fraction of each weed's potential. Of course, if you have a serious medical condition, or are already using other medicines, always seek advice from a professional before using a plant medicinally.

Amaranth

Amaranthus species

Green amaranth with flowers

The amaranths have had a long and colourful entanglement with human cultures. They have been valued for their nourishing seeds and leaves, adored for their flowers, and both worshipped and maligned. Several species grow wild in

Australia, and one of the most common is green amaranth (*A. viridis*), but it is possible to encounter any of the varieties mentioned at the end of this section.

Green amaranth leaves can contain an astounding 38% protein by dry weight, and are store to an embarrassment of mineral nutrients. Indeed, along with dandelion, it's one of the five most nutritious vegetables ever tested by the US Department of Agriculture. Both leaves and seeds contain the essential amino acid lysine, which is lacking in wheat and other 'true' grains, making amaranth an excellent nutritional partner to wheat and rice. Amaranth also contains oxalic acid (see our Notes of Caution in Chapter 1).

Pick both young tips and leaves from healthy, tender-looking plants. Unless *very* young they are best cooked. Amaranths are largely warm-season-active plants, and in temperate climates the greens are usually best from spring to mid-summer.

Amaranth is an all-purpose cooking green with a most agreeable flavour. In Greece it is served as a simple side dish: steamed or boiled then

Green amaranth illustration from *Botanischer Bilder-Atlas* (1884)

dressed with lemon juice and olive oil, often to accompany fish. The tiny poppy-like seeds are available in supermarkets puffed as a breakfast food. Some amaranth species can bear up to half a

million seeds per plant, with seed heads of a kilogram in weight. We've ground and mixed them into breads and pancakes for a delicious nutty flavour, but winnowing them is a delicate and time-consuming process.

Because of amaranth's drought tolerance, high nutritional quality and lavish seed production, this ancient plant has many

HISTORY AND CULTURE Ancient Greeks considered love-lies-bleeding (*Amaranthus caudatus*) to be sacred and decorated tombs with it. However, the earliest archaeological record of amaranth cultivation comes from Tehuacan Puebla, Mexico, dated around 4000 BC, making it one of the oldest known food crops. The seeds were one of the staple foods of Aztecs and Incas, the latter mixing it with honey or cactus syrup – and apparently even human blood – before forming it into sticky idols for communion with the gods.

A similar mix of the puffed seeds with honey, molasses or chocolate – but minus the blood – is sold in Mexico today, and called *alegría* (meaning joy or jubilation).

times been hailed as a 'grain for the future'. Recent news reports, however, are just as likely to call it 'evil pigweed', for an amaranth species in the US has developed resistance to the herbicide glyphosate (as found in Roundup®), and is smothering genetically modified cotton and soya crops, to the great distress of the Monsanto corporation. This plant, which was once suppressed by Spanish colonists for its associations with pagan rituals, seems to be rising again like an ancient god to smite the crops that supplanted it!

LOOK FOR   Smooth upright stems with alternate, vaguely diamond-shaped, dull green leaves that are rough to touch, with very apparent indented veins. The branches terminate in long cylindrical seeding spikes of various colours, depending on species – green amaranth starts out green and turns brown as the seeds ripen, while other species have red flowers of varying shades.

DISTRIBUTION   Amaranths are found widely in all Australian states, though green amaranth is rare in Tasmania. All species tend to prefer full sun, and are adapted to a wide range of climates. There are native amaranth species on all continents (except Antarctica).

RELATIVES YOU MIGHT RECOGNISE   The genus *Amaranthus*
includes garden varieties such as the stunning love-lies-bleeding
(*Amaranthus caudatus*), known for its audacious drooping red
flowers and massive edible seed production, and Chinese
spinach (*A. tricolor*) and redroot amaranth (*A. retroflexus*), known
for their edible leaves. There are several native amaranths
(including *A. interruptus*) which are most common in the nation's
north. They are eaten as greens and seed, and ancient remains
have been found in the Kimberley in a cave that shows signs of
being continuously occupied for over 40,000 years.

The red amaranth 'love-lies-bleeding' (*A. caudatus*)

Angled Onion

Allium triquetrum

ALSO KNOWN AS Three Cornered Leek, Three Cornered
Garlic, Triangle Onion, Onionweed

Originating from the Mediterranean, this pungent and delicious salad green emerges with the autumn rains and
grows lushly through winter. It produces charming white nodding
bell flowers before retreating into dormant garlic-like bulbs as
spring heats up.

Angled onion from *Flora Graeca* (1806)

When used raw the whole plant, including bulbs and flowers, excels as a spring onion substitute. It has a mellow, sweet, onion-garlic flavour as good as any cultivated *Allium*. Neither bulb nor greens are as good cooked, becoming less flavoursome and somewhat fibrous. To harvest the greens, cut whole handfuls off just above ground level (this eliminates the task of cleaning the earth from them), or pull them up if you wish to use the white base, or to thin your clump, thereby encouraging the remaining plants to grow larger and plumper.

Angled onion prefers a rich, moist, well-drained soil and grows well in partial shade. Wild, it is most abundant along waterways, but as this is one good-looking weed, with its verdant foliage and fairytale flowers, we choose to give it a nook in the garden to grace with its pungent presence year after year. Some people warn of this plant's rapaciousness in the garden, but we have found that as long as we vigilantly pull

A pre-flowering patch – note the strong ridge that gives the leaf its triangular cross-section.

out the handful of self-seeded plants that pop up away from that allocated nook each autumn, it is easy to tame.

Although no specific medical benefits are recorded for this species, it likely has some of the formidable qualities of garlic, which include cancer protection, reduction of cholesterol levels and powerful antibiotic effects.

LOOK FOR   Fleshy, bright green leaves to 40 cm high that, from a distance, could be mistaken for grass. The leaves have a distinctive triangular cross-section. The flower stem is even more triangular and is topped with multiple drooping small white bell-shaped flowers in winter and early spring. Could be confused with the cultivated flower snowdrops, but the oniony smell is unmistakable.

DISTRIBUTION   All states except the Northern Territory, but far more common in the southern states.

RELATIVES YOU MIGHT RECOGNISE   The genus *Allium* also includes leeks, garlic, onions, and shallots.

Blackberry

Rubus fruticosus and relatives

This is one plant that knows how to protect its wares, and no wonder, given how utterly delicious they can be. Blackberries ripen in late summer, and should be eaten once they have turned a deep lustrous black. We prefer them sun-warmed and au naturel. However, they can be cooked in crumbles or a zillion other fruity desserts, or made into jam or cordial. If you have access to a good patch it is easy to pick enough fruit over a season to try every blackberry recipe that tickles your fancy.

Blackberries are also somewhat of a nutritional superstar of the fruit kingdom. They are high in vitamins C and K, manganese and folic acid. But they are most notable for their

antioxidants. One US study put them at the very top of a list of 1000 foods ranked by the antioxidant content of an average serving size.

There is technique to blackberry picking, and it's not all about wearing full motorcycle leathers. To get to those juicy

Fruit are often borne at several stages of ripeness simultaneously.

berries that lie deeper within the spiny thicket, you must stomp down the highest canes in your path with your foot, using them to flatten to the ground all the canes beneath them. Do this with each step into the bushes' centre, and you'll quickly triple your collecting area. Fruit collected very late in summer may have become home to various insects or moulds, so it is worth being a tad more observant as the season wanes.

Blackberry illustration from *Kräuterbuch* (1914)

The tannin-rich leaves and roots have been valued medicinally since at least the time of the ancient Greeks. Traditionally they are made into teas whose astringency is used to relieve sore throats and mouth ulcers, as well as diarrhoea and thrush. The leaves have powerful antibacterial properties.

LOOK FOR   The plants are mostly deciduous in winter. The leaves resemble those of its relative, the rose, and are a dark matte green with finely serrated edges and paler undersides. The plant's thorny brambles grow in clumps, which enlarge as the tips of canes bend down, touch the ground, and take root. In spring, small, five-petalled white flowers appear, maturing into berries that are hard and green to start with, then bright red, then the deep purple black that tells your taste buds to get ready for a very nice time.

DISTRIBUTION   A European native, it is found in all Australian states except the Northern Territory. They prefer cooler climates and are mostly restricted to areas south of Brisbane on the east coast, and Perth on the west. Favoured locales include gullies, waterways, paddocks and vacant blocks.

RELATIVES YOU MIGHT RECOGNISE   The genus *Rubus* also includes raspberries, loganberries (a blackberry-raspberry hybrid), cloudberries and dewberries. Blackberries are in the family Rosaceae, which also includes apples, strawberries, roses, almonds, quinces, peaches, pears, plums and apricots.

NOT-SO-SWEET REVENGE   The fact that late-season blackberries can host less-than-tasty moulds was once reinforced in England by folklore. It was said that when expelled from Heaven, Lucifer fell square into a blackberry patch. So, each year on Old Michaelmas Day in early autumn, the devil curses blackberries, spoiling the fruit – by urinating on them!

Blackberry Nightshade
Solanum nigrum

ALSO KNOWN AS Black Nightshade

This plant is widely and incorrectly referred to as deadly nightshade in Australia, a misconception so common that we have had horrified passers-by try to 'rescue' us from eating them when harvesting in public places. (There *is* a somewhat similar

Unripe green berries, flowers and ripe black berries

looking plant which should be more properly called deadly night-shade: *Atropa belladonna*, but it is not naturalised in Australia.)

Blackberry nightshade, a Eurasian native, was introduced into Australia as a vegetable during the goldrush. The fully ripened black berries have a rich flavour, sweet but with savoury hints of their cousin, the tomato. They can be mixed with other fruits as a dessert, provide a sweet-tangy element in a salad, and make a fabulous addition to chutney. Early settlers' cookbooks mention using them as a pie filling but, although it is easy enough to pluck a few handfuls as a daily treat, you'd need some patience (or some early settlers' children at a loose end) to gather enough for such an enterprise. The leaves and tender shoots are eaten widely across the globe, but some parts of the plant can be toxic, particularly the green berries, so please read the cautionary information on the following pages, before use.

Blackberry nightshade is a formidable medicinal plant, with mention of it dating back to the earliest herbals. Dioscorides (circa 40-90 AD), in texts that were influential for 1500 years, recommended its leaves for treating skin diseases, earaches, indigestion, and internal bleeding. *Gerard's Herbal* of 1636 reported it to be good for ulcers, ringworm, shingles and 'panic of the head'.

Chinese medicine uses juice from the leaves against the pain of kidney and bladder inflammations, and also to remedy heartburn. In Africa, the plant is widely used for several complaints, with reported successes for conjunctivitis and ulcers. Indian Ayurvedic tradition has the plant's leaves heated and applied to swollen testicles … among other uses.

Research has confirmed its anti-herpes properties, and author and herbalist Pat Collins prescribes an ointment made from the entire plant for cold sores. Various experiments suggest its value in suppressing cancerous tumours. It has proven abilities to heal gastric ulcers, prevent epileptic seizures, and act as an anti-inflammatory. All in all, an exceptional plant.

CAUTION   Although both the leaves and berries of blackberry nightshade are eaten around the world by hundreds of millions of people, from Nepal to Hawaii to Madagascar, levels of toxins may vary regionally. Keep your intake moderate, and when choosing fruit, stick to the fully ripened ones: these are completely black, drop easily into your hand with a gentle tug, and have no bitter flavour.

There are actually several plants, including natives, in the '*Solanum nigrum* complex': plants so similar they can be difficult to tell apart. The two most common are glossy nightshade (*S. americanum*) and velvet nightshade (*S. chenopodioides*), the former being native to both Australia and the Americas. Glossy nightshade has, as the name suggests, glossy berries, which are edible when

A blackberry nightshade illustration from a 6ᵗʰ-Century version of *Vienna Dioscorides*

fully ripe and taste particularly good. Much less is known about velvet nightshade, except that the small matt berries of this velvety narrow-leafed plant are made into jam in South Africa.

Although the leaves of glossy and blackberry nightshades make excellent pot-herbs, limiting your intake is advisable. In parts of the world where they are a mainstay vegetable,

Plants often exhibit a purple tinge.

processing often includes boiling and changing the water. Solanine and its related compounds (the green potato toxins) may be present in the leaves to varying degrees, and are only destroyed by very hot cooking such as frying. Let your tongue guide you here, as their bitter taste should be very apparent.

Toxic alkaloids may be responsible for some of the tumour-fighting properties, so in this case a hint of bitterness may actually be welcome, though – as with green potatoes – not if you're pregnant.

LOOK FOR   A branching plant growing to an average height of one metre, with fairly upright stems and mildly jagged teardrop-shaped leaves of a dark or even purplish green. By mid-summer the growing tips are bearing groups of small white starry flowers that become dangling clusters of berries – initially green then ripening to a matt black.

DISTRIBUTION   Grows in all states, on almost any soil type, and usually in full sun.

RELATIVES YOU MIGHT RECOGNISE   The genus *Solanum* also includes tomato, eggplant, potato, tamarillo and pepino.

Chickweed
Stellaria media

This is a delicate annual that grows commonly in veggie garden beds, pot plants, and wherever there is moist, rich soil. A European native, chickweed tastes, in our opinion, a little like grass, albeit pleasant grass. Some more charitable types declare it delicious, 'like young corn'. Either way, it is abundant in backyards during the cooler months, and is very good for you indeed, making it a 'superfood' and a convenience food all at once.

Chickweed (*bottom*) with petty spurge (*top*). Don't confuse them! (See also page 128.)

It is loved by chickens, which lend it their name. And they are onto a good thing, as chickweed is not only high in protein, but has more than twice the iron levels of spinach, and is high in vitamins A and C and anti-ageing antioxidants.

We use chickweed raw in salads or sandwiches, but chopped finely because of the strong, springy fibres in the stem. Its delicate flavour comes into its own when paired with a more pungent salad green, such as nasturtium leaves. It can also be cooked (although it will substantially reduce in volume) or made into pesto.

Chickweed illustration from *Taschenbuch zum Pflanzenbestimmen* (1918)

Choose lush thick stands of the plant, and harvest with scissors by snipping off the top three to five centimetres (including any flowers), as if you were trimming a beard. This method will give you more leaf and less stem, and will inspire your chickweed patch to grow a new set of lush leafy top shoots.

The classical Greeks and Romans sometimes cultivated chickweed, and it was a famine food throughout Europe until recent

SALAD SALON   The Ainu people are an indigenous tribe of what is now Hokkaido, Japan. According to Ainu mythology, the first humans had bodies made of earth, spines made of willow sticks, and hair made of chickweed!

times. It was popular in Japan in ancient times, and it is still eaten in that country today as part of a symbolic seven-herb rice porridge meal. This mid-winter dish, called *nanakusa-gayu*, is eaten to promote longevity and health.

Like some other foodstuffs (including olive oil, red wine and soybeans), chickweed contains saponins: natural soaps which may lower cholesterol, but can be toxic in extreme doses. Fortunately only tiny amounts are absorbed, with the remainder passing through us harmlessly.

Chickweed has a long tradition of medicinal usage, most often for skin complaints, especially those that involve itching. Apply as a poultice, by crushing the fresh plant and securing it to the afflicted area with plastic wrap, or make an ointment by simmering a good quantity of the plant in olive oil, straining off the oil and combining it with beeswax.

LOOK FOR   A bright, fresh-green plant forming a low mat, 5-25 cm high. It has delicate stems with a strangely elastic core, small, tender teardrop-shaped leaves, and tiny star-shaped white flowers from late winter. The key identification tip is to look for a line of tiny hairs, like a mohawk, along the stem between nodes. Common look-alikes include petty spurge (*Euphorbia peplus*), which you certainly wouldn't want to eat (see Chapter 3, Other Weeds). It has a bluer tinge and produces an irritant white sap.

DISTRIBUTION   Found in all states, though rare in the tropics. Prefers very fertile soil, and grows happily in full sun or partial shade.

Distinctive 'mohawk' of hair on chickweed stem

Cleavers

Galium aparine

ALSO KNOWN AS Sticky Weed, Goosegrass

A plant familiar to many people since childhood, as the whole plant is covered in fine hooks that endow it with the mischief-inviting ability to stick to clothing. Cleavers has edible seeds, stalks and leaves, and the taste is gentle and pleasant, but with the obvious drawback that the texture is not unlike Velcro. As a steamed vegetable it's tolerable, as the hairs soften, but far better is to blend it thoroughly. Our most common use for it is in a green smoothie, and it can contribute to a fine blended soup.

Perfect for picking!

Native to North America and Eurasia, cleavers has a centuries-long tradition of use by herbalists, most commonly for burns and skin complaints, but also as a mild diuretic and for cleansing the lymphatic system. Lymphatic flow stimulators help clear cellulite and, indeed, cleavers' reputation on this front goes back to ancient times. Pliny the Elder reported: 'A pottage made of Cleavers, a little mutton and oatmeal is good to cause lankness and keepe from fatnesse.'

It has been used as an antiperspirant by the Chinese, and by milkers to strain out animal hair from milk in Sweden

Cleavers illustration from *Flora von Deutschland Österreich und der Schweiz* (1885)

Mature cleavers with its pairs of tiny fuzzy seeds

and elsewhere. In Turkey *Galium* species are called *yoğurt otu*, literally 'yoghurt herb', because they contain enzymes that can coagulate milk. Chickens love the seeds, and its other name, goosegrass, refers to the epicurean esteem in which it is held by geese. A red dye can be obtained from a decoction of the root and, when ingested, can dye bones red. We struggle to think of any useful application for that, so we'll stick to enjoying cleavers in smoothies.

LOOK FOR   This is a bright green, sprawling plant, with multiple long stems that climb and trail over the ground and other plants, climbing not much more than a metre. The narrow leaves

The miniscule hooks that give cleavers its clingy properties.

are arranged in whorls of 6-9 around the stems and are covered with fine hook-tipped hairs, as are the stems. In spring to early summer, tiny green-white flowers appear, followed by pairs of seeds the size of match-heads, also covered in fine hooked hairs.

DISTRIBUTION   A common garden weed, it can be found in all states, but is rare in the tropics. It prefers moist, fertile soils and can grow in full sun or shade.

Dandelion

Taraxacum officinale

A flowering patch of dandelions growing in good conditions

With its toothed leaves, cheerful yellow flowers and gossamer ball of parachuted seeds, dandelion is arguably the most iconic of all weeds. It has also been a much-savoured vegetable since antiquity, and valued for its curative properties. It's a bitter green, but one well worth developing a fondness for; it is not only plentiful, but highly nutritious. Indeed, unless we were to count parsley, which is usually considered a herb, it is the most nutritious vegetable ever tested by the US Department of Agriculture! Dandelion is particularly high in iron, calcium, vitamins A, B6, E, and K, thiamin, antioxidants and beta- and alpha-carotene, consumption of the last being associated with a long and healthy life.

All parts are edible. Choose young green leaves from near the centre of the plant, and ideally from plants that are yet to flower. If you are sensitive to bitterness, salt helps to 'neutralise' this flavour for your taste buds, and you can also chop dandelion finely and combine it with milder ingredients. Cooked, the leaves complement the sweetness of root vegetables and oniony fry-ups, and aid in the digestion of fatty meats. The flower petals of dandelions are sweetish and tender, and can be torn out and added to omelettes, patties, sandwiches or salads.

The roots are eaten as a vegetable in Japan, but there may be technique involved, as we've found them pretty tough. They do make for a tasty and popular coffee substitute though. To prepare, harvest roots in autumn, cut into chunks and roast in a slow oven until dark brown and brittle. Grind in a coffee grinder, and steep the grounds with boiling water, adding spices, milk or honey as you wish. Dandelion is a short-lived perennial, and while the best greens come from younger plants, the bigger roots come from the larger, more veteran specimens.

While the common name dandelion comes from the ancient Norman *dent-de-lion*, or lion's tooth (referring to the shape of the leaves, and perhaps also to the yellow lion's mane of its flowers), this esteemed plant's botanical name denotes other attributes. *Taraxacum* is thought to be from the old Greek words *taraxos* (disorder) and *akos* (remedy), and to this day dandelion is used all around the world as a medicinal herb. Most often it is applied in the treatment of liver diseases, kidney and spleen complaints, skin conditions and dyspepsia, and as an appetite stimulant

Dandelion illustration from *Köhler's Medizinal-Pflanzen in Naturgetreuen Abbildungen mit Kurz Erläuterndem Texte* (1887)

and digestive aid. In both Mexico and Turkey it is used as an anti-diabetes medicine. Early studies suggest dandelion may be useful in fighting breast and prostate cancer, chemo-resistant melanomas and leukaemia.

Dandelion has also been used in traditional Chinese recipes against acne, and an extract of the leaves was demonstrated to suppress the microbial activity that causes breakouts. Nobly using ourselves as guinea pigs, we have also tested the folk remedy of breaking the plant's flower stalk and dabbing the latex on any stray pimples, and found them looking distinctly more sheepish by the next morning.

Humans have been spreading dandelion, both for cultivation and by accident, for so long that while it is believed to be originally from the Mediterranean, its exact origins are unclear and it is now considered native to most of the Northern Hemisphere. LOOK FOR   Rich mid-green leaves growing in a rosette (that is, from a central point at ground level). The shape of the leaves varies, with margins ranging from deeply toothed to only slightly serrated. From the centre of this rosette emerges the hollow flowering stalk, which bears a single bright yellow flower with densely layered petals. The flower transforms into the pale sphere of downy seeds that can be blown away by wind, or by human breath while being wished upon. The root is a taproot shaped like a slender white carrot. All parts of the plant emit a white latex when damaged.

Dandelion closely resembles the common cat's-ears (*Hypo-chaeris radicata* and *H. glabra*), which are slightly tougher and hairier looking, usually have multiple flowers on each stalk, and do not produce a white sap. In any case, cat's-ears are also perfectly edible. Dandelion could also be confused with the native murnong or yam daisy (*Microseris lanceolata*), whose

Dandelion leaf variations

tubers were once a staple carbohydrate for people in Victoria but sadly now rarely exists wild. Overgrazing by sheep and soil compaction are blamed for its demise across the State. The dandelion, with its compaction-curing taproot, is now one of the plants helping to reverse this degraded state of many pastoral soils.

DISTRIBUTION   Dandelion can be found in all states of Australia, but is less common in the tropics. It will pop up in lawns, but prefers less trodden areas where it mingles happily with lush grass at the edges of parks and pathways, or in pastures.

Fat Hen

Chenopodium album

ALSO KNOWN AS Lambsquarters, Goosefoot, Wild Spinach

Fat hen is a summer green that grows large and lustrous in the garden. Of all the wild greens covered in this book, fat hen may be the one most interchangeable with its domesticated cousin spinach. We have received confessions from previous weed walk participants of sneaking it into dinners served to their less 'food-flexible' partners, with the only comment received being that that dish was particularly delicious that night. The flavour is mild with a slight nuttiness, the texture is silky, and it is a breeze to harvest.

Typical growth habit

While fat hen can handle dry soils and intense heat, the best picking comes from healthy looking plants on rich or cultivated soil. We generally harvest just the young leaves and growing tips. The greens are best cooked, due to high oxalic acid content

THE TRUTH ABOUT MELBOURNE   The old English name for fat hen is 'Melde'. With that in mind, we bring you this newsflash from *The Age* newspaper (6 September 1965):

> INFORMATION on the origin of the name Melbourne has been received – rather distressing information. This fine city it appears was named after a weed whose Anglo-Saxon name was Melde (*Chenopodium album*).
>
> The Melbourne family of Derbyshire and Cambridge, whose member Lord Melbourne gave his name to the Australian city, is itself named after the weed. Until AD 970, the name was spelt Melde-Bourne because at one time large quantities of Melde were grown in these two counties for food and fodder.

Pick fat hen leaves young

By the time seeds are
formed, the greens are
less palatable.

(see our Note of Caution in Chapter 1). Steamed and served
simply with olive oil, lemon juice and salt, they are superb. Fat
hen is widely grown in Northern India, and used in many dishes
including the popular *bathue ka raita*, in which it is cooked and
mashed with yoghurt and spices. Fat hen also gives spinach a
run for its money in the nutritional stakes, and comes out trumps
in many departments. It is particularly rich in vitamin C, ribo-
flavin, calcium and antioxidants, and has been tested at up to
43% protein by dry weight. Body builders take note!

Illustration of fat hen from *Deutschlands Flora in Abbildungen* (1796)

The seeds too are sometimes harvested and eaten – since at least the 4th century BC in fact: fat hen seeds were among the foods in the stomach contents of the amazingly well-preserved Tollund Man found in a Danish peatbog. Fat hen has been successfully grown from seeds found in undisturbed soil at another Danish archaeological dig; seeds believed to be 1700 years old!

Fat hen is celebrated for its nutritional qualities more so than its medicinal ones, but research has validated a traditional Pakistani use as an anti-parasitic against intestinal worms and as an anti-inflammatory.

LOOK FOR An upright central stalk of between 30 cm and 1.5 m bears side branches with jagged-edged leaves of a dull darkish green, often with a blue tinge. The underside of the leaves and the stem at the growing tip are coated with a waxy white powder. The green bud-clusters appear later in summer, turn red-brown as they mature, and bear tiny shiny black seeds. It could be mistaken for its cousin orache (*Atriplex prostrata*), which tends to grow near water, has more lance-shaped leaves, and is also edible. It has a close relative which looks quite similar except that its leaves are a more vibrant green and don't have white undersides. This is known as green fat hen (*C. murale*), and is eaten in the same ways.

DISTRIBUTION Fat hen is a European native, but well adapted to many conditions and available across all states of Australia, usually in sunny, disturbed areas.

RELATIVES YOU MIGHT RECOGNISE The genus *Chenopodium* also includes quinoa (*Chenopodium quinoa*) and the beloved Mexican vegetable *huazontle* (*C. nuttalliae*).

Fennel

Foeniculum vulgare

Fennel's characteristic branching habit

Fennel, a Mediterranean native, has been cultivated since the times of the ancients for its aromatic aniseed flavour and therapeutic qualities. Its tufts of delicate foliage often decorate sloping ground, and it seems to have a special love for the embankments of railway lines.

A beneficial insects' utopia

Try the fresh green and yellow seeds for a sweet breath-freshening sensation that reputedly doubles as an appetite suppressant. Alternatively, let the seeds dry and grind them with a mortar and pestle for use in recipes (perfect in slow-cooked stews) or for making a sweet tea. Pick the very young foliage – you're looking for the stuff that is almost fluorescent green – and chop finely to use as a herb. Classic culinary tradition often finds it paired with fish or chicken, or used in salads and cream sauces.

The wild varieties of fennel do not create a bulb like some of the cultivated ones, but they do make it possible for ordinary humans with ordinary bank balances to sample the gourmet ingredient of the moment: fennel pollen. This gold dust is

Fennel illustration from Köhler's *Medizinal-Pflanzen in Naturgetreuen Abbildungen mit Kurz Erläuterndem Texte* (1887)

gathered by swanning about shaking the flower heads of lots of fennel plants into a paper bag. You then fold it into the dough for home-made pasta, dust it over cream-tossed, wild pine mushrooms in a verjus reduction, roll cubes of goat's cheese in it and arrange them on ridiculously large white plates, etc. ...

The seeds are predominant in the sugar-coated Indian seed mix *mukhwas*, eaten as a digestive after meals. Gerard reported in the early 17th century that, 'Fennell seed drunke asswageth the paine of the stomacke ... and desire ... to breake winde,' the latter making them quite strategic after a legume-filled curry. Fennel is an anti-spasmodic herb for the digestive system. It may be for this reason that it is part of an effective herbal mixture that helps colicky (incessantly crying) babies. An oil is extracted from the seeds and, traditionally, this or the seeds have been used to treat chest colds. The oil also has powerful antifungal and antioxidant properties.

Not only do fennel's arching fronds make it very ornamental in the garden, but it also assists in natural pest control (see below, under 'Look for'). However, its clump expands readily, so give it a (sunny) spot in your garden by all means, but a little apart from less vigorous plants. It lives for several years, but withers somewhat in winter, which is a good time to chop it back hard to encourage lush new spring growth.

LOOK FOR   Tufts of feathery foliage that grow on stout, upright stalks. The leaves are as fine as thread, and range from vivid light green (very young) to a bold dark green (older). As with most other members of its botanical family, fennel bears its many tiny yellow flowers in an umbrella-shaped spray; a shape that is particularly inviting to many beneficial predatory insects, including ladybirds, lacewings, and the various parasitic wasps

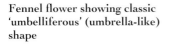

Young fennel good for picking

Fennel flower showing classic 'umbelliferous' (umbrella-like) shape

that attack everything from aphids to codling moth. Flowers become green seeds that mature to a pale silvery brown as summer concludes.

Be aware of the potential look-alike, toxic hemlock (see our Note of Caution in Chapter 1). The resemblance is not great – hemlock does have lacy leaves but they are much more fern-like and not nearly as fine as those of fennel. For this reason, never harvest seeds from plants with dead foliage. Hemlock also has none of fennel's characteristic aniseed smell.

DISTRIBUTION   Found from the sub-tropics to cool temperate regions, in all states except the Northern Territory.

RELATIVES YOU MIGHT RECOGNISE   Fennel is in the family *Apiaceae*, which also includes cultivated bulb fennel, dill, caraway, coriander, parsley, celery, carrot and parsnip.

Mallow

Malva parviflora, *Malva neglecta* and
similar *Malva* species

The ancient Romans considered mallow a delicacy, and to this day mallow species (both wild-harvested and cultivated) are widely eaten throughout the Mediterranean, Middle East, Northern Africa and China. Mallow grows year round, has a pleasant and mild flavour, and is consistently one of the most popular on our weed walks. Mallow shares a little of the ability of its relatives okra and marshmallow to thicken the dishes it is cooked in, known as 'mucilaginous' properties.

Mallow's young growth makes for the best eating.

Use the young leaves only (the older ones get a bit leathery), particularly if using them raw in salads. Cooked, it can stand in for spinach in any recipe, although it has more body, plus that slightly gooey texture which many cultures adore.

The seed heads (sometimes called 'mallow cheeses', probably because of their cute 'wheel of cheese' appearance) can be harvested quickly in respectable quantities when you find a large stand. Choose the pale green ones that haven't yet become dry and brown, and shop around, as some species' seeds are a bit fibrous. Add them to risotto or curries, lightly steam and dress with olive oil and lemon, or fry with butter, onion and mushrooms. These seed heads are high in carbohydrates, oils and proteins: a potential staple food.

Mallow is one of the earliest plants mentioned in recorded literature. In 30 BC the Roman poet Horace wrote: *'Me pascunt*

Illustration of small mallow (*Malva neglecta*) from
Deutschlands Flora in Abbildungen (1796)

olivae, me cichorea, levesque malvae.' (I graze on olives, chicory and simple mallow.) He believed that mallow 'develops the intellectual faculties'. Pliny claimed that, 'whosoever shall take a spoonful of the mallows shall that day be free from all maladies,' and to this day there is a saying in southern Italy, *La malva, da ogni male ti salva.* (Mallow saves you from every disease.) Dioscorides recommended mallow for the treatment of burns and skin inflammations, spider bites, bee and wasp stings, bowel and urinary problems – uses that have persisted into modern times.

One recommendation that hasn't survived: 'Applied with urine it cures running sores on the head and dandruff.' (Check out our nettle profile for a dandruff cure that doesn't require you to put pee on your head.)

These days, herbalists utilise mallow's slimy mucilage for coughs, and to soothe inflamed throats and digestive tracts. A recent study found an extract from tall mallow (*M. sylvestris*) to be more effective than cimetidine, a potent drug for treating gastric ulcers.

In the garden, its deep taproot is one of nature's most talented at penetrating hard clays, leaving behind root tunnels that allow worms, air and water to begin moving through these problematic soils. Employ a few generations of mallow to start reclaiming that solid, sticky clay area in your garden.

LOOK FOR   Darkish-green rounded leaves, with slightly paler undersides, wavy margins, and veins that radiate out from their long stalks. They are fuzzy to the touch and look vaguely like the leaves of a geranium, but are less fleshy. The white to pale-purple flowers have five petals and are followed by the rounded seed capsules - a milky green maturing to a golden brown and partially encased by what was previously the flower calyx.

Mallow flowers

Mallow seeds, most likely
M. linnae

DISTRIBUTION   *Malva* species are
natives to Eurasia and Northern Africa.
They can be found in all states, and
anywhere from lawns to dry embankments to moist riverbanks.

RELATIVES YOU MIGHT RECOGNISE
Mallow is in the family Malvaceae,
which also includes marshmallow, okra,
hibiscus and hollyhock. There are at
least ten common weeds in the *Malva*
genus in Australia, which also includes
a native known as Australian hollyhock
(*Malva preissiana*). All are considered
edible.

Nasturtium

Tropaeolum majus

This South American native is beloved by many gardeners for its fiery flowers and easy cultivation. Leaves, flowers and seeds are all edible, and all share a sweet peppery flavour that is acknowledged in the name 'nasturtium', which has its roots in the Latin for 'contorted nose'. The dried ground seeds were even used as a pepper substitute in World War II.

The leaves are delicious in salads, can be used like vine leaves for making dolmades, and make an excellent watercress

Nasturtium illustration from *Afbeeldingen der artseny-
gewassen met derzelver Nederduitsche en Latynsche
beschryvingen* (1796)

substitute in egg sandwiches. As kids we liked to bite off the 'tail' at the back of the flower and suck out the nectar, but grown-ups can enjoy the flowers too: they look amazing on top of a salad, can be stuffed with cream-cheese mixtures if you want to make hors d'oeuvres like it's 1974, and unopened flower buds and seeds can be pickled and used like capers (see our recipes in Chapter 4). It really works! Nasturtiums' spiciness is more pronounced when the plants are growing in heat and sun.

Those spicy mustard compounds can also take credit for many of nasturtiums' medicinal properties. The plant has been used in Andean herbal medicine as a disinfectant, wound healer, antibiotic, and an extract has been tested as having effective anti-inflammatory and selective antimicrobial properties. Mexican herbalists use a topical application of nasturtium for ringworm, and also claim the plant to be good against cancer of the left lung! They might like to include the right lung too: benzyl mustard oil, which can be extracted from nasturtiums, has been shown as effective against several types of tumorous cancers when taken orally. These mustard oils are also known fungicides, and herbalists have prescribed soaking in a nasturtium bath for cases of athlete's foot. Nasturtiums, especially the flowers, are extraordinarily rich in the yellow pigment lutein, which is important for eye health and preventing blindness. Studies on nasturtium tea have found it to be a powerful diuretic and treatment for high blood pressure, and a blend of horseradish root and nasturtium is a proven treatment for urinary tract infections.

In the garden nasturtium is considered a good companion plant and is often planted at the base of apple and pear trees to deter codling moth.

Nasturtium flowers come in shades from pale yellow to crimson red, or can have bicolour petals within this spectrum.

LOOK FOR A trailing, rambling, or climbing plant with lush mid-green, coaster-sized round leaves held aloft on long stalks which attach to the leaf's centre. The large gaudy flowers are orange, yellow, or sometimes red, and feature a long hollow 'tail' filled with nectar. The seeds are pale-green pea-sized grooved nuggets, often borne in trios.

DISTRIBUTION Often a garden weed or garden escapee, nasturtiums can be found in all states along watercourses, road-sides and urban bush. It is seldom found wild in the tropics.

RELATIVES YOU MIGHT RECOGNISE You would be forgiven for thinking that nasturtiums are related to watercress, what with watercress being a member of the genus *Nasturtium*. But they aren't even in the same family, let alone genus. Nasturtiums' common name was bestowed due to its peppery flavour being so similar to that of the cresses of the *Nasturtium* genus.

Nettle
Urtica urens

ALSO KNOWN AS Dwarf Nettle, English Nettle, Stinging Nettle

Of all the weeds in this book, nettle is the easiest to identify – you can do it with your eyes closed! For anyone who has ever been stung by this plant, the spice that is revenge can only add to your enjoyment of having nettles for supper. Nettles are one of the most nutritious and versatile greens available: high in powerful antioxidants, and bountiful in protein – up to 40% by

A patch of young nettles

dry weight. They are dense in minerals – almost extraordinarily so in the case of calcium; one modest serving of around 150 g would satisfy your recommended daily intake.

Nettles' feisty sting comes courtesy of tiny silica hairs filled with a mischievous cocktail, including formic acid, the substance also responsible for the persuasiveness of ant stings. The easiest way to pick and prepare your nettles is by wearing a glove on one hand to hold the stalk, while wielding scissors or a knife in the other.

Collect only young leaves because older leaves may develop gritty particles called cystoliths, which irritate the kidneys. The stems are highly fibrous, so take only the soft top of the stem for eating. If you do get stung, the crushed or chewed leaves of dock, which happily often grows nearby, is a traditional anti-dote, championed in the old English rhyme:

> *Nettle out, dock in –*
> *Dock remove the nettle sting.*

Blending, drying and cooking all disarm the sting. You can simply blanch the leaves in boiling water for a minute, then plunge them briefly into cold water to preserve the electric green colour. Then you can use them without fear in your chosen dish. Nettle culinary classics, issuing from a long Greek and Italian affection for the plant, include nettle gnocchi with sage butter (for our recipe, see Chapter 4), nettle and ricotta ravioli, nettle soup, nettle spanakopita and nettle pesto. Simple nettle tea is regarded as 'strengthening', and tastes both fresh and earthy, with slight seaweed flavours. Steep fresh or dried leaves in boiling water, or simply reserve the water strained off from cooking.

Nettle fibres have often been used to make fabric, and the whole plant to make dye; the leaves are such a vivid green that it's easy to see why. This rich chlorophyll-saturated hue

A nettle just at the point of seed development

also lends easy credence to their reputation as a blood tonic, and baskets of the plants were once hawked in the streets of London to the call of 'Nettles with tender shoots, to cleanse the blood!' Despite this enthusiasm, the poet Thomas Campbell

Illustration of nettle from *Deutschlands Flora in Abbildungen* (1796)

was disappointed by the neglect of nettles in England and wrote, 'In Scotland, I have eaten nettles, I have slept in nettle sheets, and I have dined off a nettle tablecloth. The young and tender nettle is an excellent potherb.'

Regular ingestion of this 'excellent potherb' can reduce the pain of rheumatoid and osteo-arthritis, but external application of the plant is the more notorious mode of treating sore joints and arthritis – self-flagellation with nettles is a practice that dates back at least to Roman times. The verb is to *urticate*; to whip nettle leaves on afflicted areas for proven symptom relief! Nettle also has proven action against enlargement of the prostate, sufferers of which will be relieved to hear involves the gentler prescription of a (commercially available) extract of nettle root. Perhaps counter-intuitively, nettle also contains antihistamines, and consuming the leaves has been confirmed as a mild remedy for hay fever.

For a hair-softening and anti-dandruff tonic, steep the chopped leaves in apple cider vinegar for two weeks, and strain. Massage a tablespoon of the liquid through wet hair for a minute or so while in the shower, then rinse and wear those black polo necks with confidence once more.

LOOK FOR A plant of about 30 cm tall, modestly branching and upright. Leaves are a 1 to 4-cm long teardrop shape and dark forest green with a serrated edge. Green seed clusters appear at the growing tip and along the stem.

DISTRIBUTION   Native to Europe and Northern Africa, nettle can be found in all states of Australia, preferring loose fertile soils and full sun to partial shade.

RELATIVES YOU MIGHT RECOGNISE   The main weedy nettle in Australia, and the one we've been talking about, *Urtica urens*, is a cool-season annual. It has a more delicate appearance and darker leaves than the rarer perennial tall nettle (*U. dioica*) and the native bush-tucker plant, scrub nettle (*U. incisa*). The genus *Urtica* also includes the tree nettle (*U. ferox*), exclusively found in New Zealand, which has such a ferocious sting it has killed dogs, horses, and at least one person.

Pale-flowered wood sorrel (*O. incarnata*)

Oxalis

Oxalis species

School children enjoy chewing on the long flower stems of some varieties of oxalis for the shock of sour juice – a refreshing taste that comes courtesy of the unusually high levels of oxalic acid (see our Note of Caution in Chapter 1) contained by members of this genus, along with citric and tartaric acid. In the time of Henry VIII, the English held oxalis in great repute

as a culinary herb, but it lost favour after the introduction of the larger, leafier, French sorrel (to which it is not related). It is rich in vitamin C, and has been used to treat scurvy. We use oxalis not as a vegetable, but as a herb. This is due partly to its intense flavour, but more to the sheer difficulty of gathering a large quantity. The leaves shrink like Alice in Wonderland when cooked, but they make a perfect tart addition to a spanakopita or an omelette or rich stew, thrown in at the last minute.

Gerard wrote, 'Greene Sauce is good for them that have sicke and feeble stomaches … and of all Sauces, Sorrel is the best, not only in virtue, but also in pleasantness of his taste.' But it is in India where the plants have been most widely used medicinally. The yellow-flowered creeping wood sorrel (*Oxalis corniculata*) has at least eighty recorded local names there! It is considered

Creeping wood sorrel (*O. corniculata*) with flowers and seed pods

Creeping wood sorrel
(*Oxalis corniculata*)
illustration from
*Deutschlands Flora in
Abbildungen* (1796)

One of the many purple-flowering oxalis species

helpful in treating conditions such as influenza, urinary tract infections, diarrhoea, sprains, insect bites, open wounds, burns and hookworm.

A number of these uses, including oxalis' ability to speed up wound healing (through an oral dose), have been backed up by testing. Oxalis species have also demonstrated anti-cancer, cardio-relaxant, and potent antioxidant properties. They have shown antibacterial activity in the lab against several human pathogens and can be used as a topical cream for skin infections.

Creeping wood sorrel was also used traditionally in India to prevent fertility in women, and an experiment on rats showed abortifacient activity. So, not a herb for women who are trying to conceive.

LOOK FOR   Low-growing (to 25 cm) clumping plants, with leaves divided into three heart shaped leaflets in a vibrant mid-green, which fold up in rain, bright sunlight, and at night. Different species bear yellow, white, or purple flowers, which mature into small capsules. Many oxalis species could be mistaken for clover (also edible, so no concerns there), but can be distinguished by their lemony taste. Incidentally, the Irish aren't concerned which one you call shamrock, as long at it has the three wee leaves.

DISTRIBUTION   Oxalis species are available throughout the continent in a wide range of climates and conditions, though those lush enough to pick usually appear in moist or semi-shaded spots.

RELATIVES YOU MIGHT RECOGNISE   The genus *Oxalis* also includes oca or New Zealand yam (*Oxalis tuberosa*), which produces a small edible tuber.

Plantain

Plantago species

There are several common species of this most useful plant, the most abundant being the strappy-leaved common plantain (*Plantago lanceolata*). In damper spots you may encounter the larger oval leaves of the greater plantain (*P. major*), and in mowed parks, lawns and drier areas you may find the small,

Common plantain
(*P. lanceolata*)

ornately formed leaves of the buckshorn plantain (*P. coronopus*).
The latter is cultivated in Italy for its mild nutty flavour, but the
leaves of wild specimens are a little too small to be rewarding.
The greater plantain variety does provide more foliage, but both
it and common plantain have a slight flavour of bitter mushrooms.

This can actually become a complementary note to other tastes, but is overbearing on its own, so choose the tenderest young leaves, and use them in a mixed salad, soup, stew, stir-fry or smoothie.

Common plantain illustration from *Flora Batava,* Volume 1 (1800)

Buckshorn plantain
illustration from
Flora Batava,
Volume 1 (1800)

Buckshorn plantain
(*P. coronupus*)

A LITTLE ETHNOBOTANY ... So closely has
the proliferation of common plantain followed
the spread of agriculture that archaeologists
look for its fossilised pollen to estimate the
expansion of early-Neolithic farming throughout Europe. Almost 10,000 years later, when
Europeans reached the Americas, greater plantain followed their movements in such a predict-
able fashion that indigenous peoples branded it
'white man's footprints'.

Psyllium powder is made from the seed husks of some cultivated varieties of plantain. These husks are abundant in soluble fibre and are used in commercial preparations (such as Metamucil) to soothe the digestive tract, improve bowel regularity, and reduce the absorption of cholesterol. You can eat the seed heads of the wild and weedy varieties chopped onto your breakfast cereal for similar effect.

The leaves are prized medicinally. Anglo Saxons called plantain the 'Mother of Herbs', and included the leaves as part of their Nine Herbs Charm, used for treating poisons, cuts and infections (both nettle and fennel also feature in the concoction). In Norway, common plantain is known as *groblad*, and on the Isle of Man it is *slan lus*, both essentially translating as 'healing herb'. Leaves and juice from the leaves are widely used on wounds of all types. They have demonstrated antibacterial and anti-inflammatory properties, and contain special bio-regulating proteins that assist in the plant's 'profound' wound healing abilities. All you have to do is chew up a leaf and *voilà*, you've made your own 'spit poultice'.

Studies have also given support to other traditional uses including powerful immune-system boosting properties and the treatment of diarrhoea, giardia, ulcers, and cancerous tumours. And, to top it all off, it has been shown to enhance physical endurance (in mice at least)!

LOOK FOR   Members of this genus are easier to recognise when in flower, as their leaves vary a great deal. Most have oval leaves (either wide and fat or long and narrow, depending on the species), growing from a central point on the ground. These are dull green, softly hairy leaves, with parallel veins, and a coarse texture. A cylindrical flowering spike that looks a little

Greater plantain (*P. major*)

like that of a miniature grasstree appears on a stalk poking up from the plant's centre, browning off as the seeds mature.

DISTRIBUTION   The three plantain species mentioned can be found in all states except the Northern Territory, and not in the tropics of any state. Usually found in full sun, often in compacted ground, it will tolerate heavily trodden areas, including pastures and parks.

Prickly Pear

Opuntia ficus-indica

ALSO KNOWN AS Indian Fig, Nopales

This plant, with its heady mix of succulent fruit, gorgeous flowers and wicked spines, is an important plant in its native Mexico, where it is used heavily as fruit, drink and vegetable. Known locally as *tuna*, the fruit are sold on the streets ready peeled and stuck three to a wooden skewer, made into jellies, candies and jams, or transformed into several brands of liquor. They have a flavour somewhere between persimmon and kiwifruit, and seem somehow extra sweet and refreshing for the unlikeliness of having come from a spiny, 'deserty' cactus.

The pads are known as *nopales*, and are eaten in tortillas, tacos, cooked with scrambled eggs or as a side with beans, rice and other dishes. They are delicately tart, and taste very much like themselves, in a way that you get a yen for once you have experienced it a few times.

If you don't have a talented Mexican cook to gather and prepare your cactus for you, don't be disheartened. Let's start with the fruit. These are egg-shaped baubles that appear on the rims of the cactus pads and mature from green to orange or a deep purplish red over the course of summer. The fruit do not have the large spines found on the pads, but they *do* have tiny hair-like spines called 'glochids', which can be very irritating if stuck in you, so you need sturdy gloves to pick and prepare your fruit. Native Americans rolled them in coarse sand to get rid of the glochids, so you could give that a go, but we generally just use a knife to slice them open, and then spoon out the brightly coloured insides. Perform this task while wearing gloves and in a spot where dropped spines can be disposed of easily (over an old plastic bag, or outdoors).

The paddles have both big naughty-looking spines and in some varieties also a modest serving of the tiny, insidious glochids, so in these cases they should also be prepared with gloves. Use a knife to trim off the rims of the paddle, and shave the spikes off its surface, then wash your utensils, chopping board and the paddle so that you can handle it like any other vegetable. Cut it into strips or dice it, and fry until it tenderises. If you want a softer texture, pre-boil the pieces in salted

Ripe prickly pear fruit

water for 5-8 minutes – this also allows you to lessen the 'slimy' factor that comes from prickly pears' sap. Many people love this okra-like texture, but if you are not one of them, you can even change the boiling water part way through, or add a pinch of bicarbonate soda for the last two minutes of boiling. *Over*cooking increases the sliminess and destroys the crispness that is part of the cactus' charm. Dress with lime juice, olive oil and crumbled feta, or sauté with other vegetables; or use as a topping with chilli and kidney beans for a brave new pizza Mexicana (see our recipe in Chapter 4). Score them halfway

A STUDY IN CRIMSON   In the 16th century prickly pears provided Mexico with its biggest export after silver, and it was nothing to do with food, but with glamour. The *Opuntia* cacti are the chosen home for the scale insect *Dactylopius coccus*, which produce a red acid in their bodies that can be made into the crimson dye cochineal. The world beyond the Aztecs had not encountered such a hue before, and it became a hugely important commodity until the arrival of the synthetic dye industry. Nowadays cochineal – also known as carmine – is mostly used in cosmetics and food colourings, where its status as a natural product is making it increasingly in demand, as health concerns about synthetic food colourings escalate.

Prickly pear (*Opuntia ficus-indica*) illustration from *The Cactaceae* (1919)

Young pads, perfect for eating

through in a criss-cross pattern, brush with oil and seasoning, and grill gently until they start to soften and char just a little. The smaller young pads picked in early spring are traditionally regarded as the best, having the least sap, mildest flavour and most tender texture.

Prickly pear fruits are rich in antioxidants, and eating them before drinking (… tequila anyone?) can also reduce the severity of a hangover. The cooked paddles have been shown to reduce blood sugar, validating their traditional Mexican use against diabetes. As in mallow, plantain and purslane, the 'slimy' element in prickly pears' texture arises from mucilage, the soluble dietary fibre that helps to sooth the digestive tract and throat.

LOOK FOR   Hard to miss this one: dull dusky green oval cactus pads or 'paddles' the size of badminton racquets covered in 1-cm spines. The pads generally cover the stem thickly, and the plant can be up to three metres tall, often forming tangled clumps that arise when old paddles drop to the ground, take root and form new plants. The flowers are large yellow, cupped, and borne in rows along the paddles' rims. By late summer they have formed spiny fruit similar in size and shape to a duck egg, and they turn orange, red or purple and are tender to squeeze when ripe.

DISTRIBUTION   *Opuntia ficus-indica* usually occurs in full sun in settled areas, across all states. Other *Opuntia* species are available across the continent, particularly in dry areas. In urban areas they are most often found along train track and creek embankments by the back fences of gardens they have escaped from.

RELATIVES YOU MIGHT RECOGNISE   You may come across many related *Opuntia* species, including *O. Stricta* which was once a widespread threat to NSW and Queensland farmers. Efforts to control it included the slaughter of many tens of thousands of emus, magpies and crows – which ate the fruit and spread its seed – until, famously, the introduction of the cactoblastis moth as a biological control. As far as we can tell no *Opuntias* are considered toxic.

Purslane

Portulaca oleracea

ALSO KNOWN AS Wild Portulaca, Pigweed, Munyeroo

Note the mat-forming tendency as the plant matures.

Purslane's glisteningly fresh little leaves leap out from bare and disturbed soil in spring. Relish it while it is available over the warmer months, for its crisp, tart succulence makes it a culinary delight. Purslane is used extensively in Middle Eastern and Mexican cuisine as both a raw and cooked vegetable. Its tangy flavour complements tomatoes, feta, roasted pumpkin, beans, fish, hard-boiled eggs … the list could go on. We harvest by plucking the growth tips (with or without flowers and seed heads) and the larger, younger leaves. Avoid picking too much stem if using it for salads, though cooking softens them nicely.

Purslane has been referred to as a 'power food', for it is the richest source of omega-3 fatty acids of any leafy greens ever tested, and is high in protein, potassium, vitamins A, C and E, and anti-aging antioxidants. It was eaten by European explorers of Australia's interior to stave off scurvy, with botanist Von Mueller declaring, 'I have reason to attribute the continuance of our health partly to the constant use of this valuable plant.'

It has been ranked as the eighth most common plant in the world, and its range is truly global. It is considered native to everywhere from North Africa to Indonesia, and while not often considered native to North America, seed deposits show it beat Columbus to the New World by at least a millennium.

As you may have gleaned from Von Mueller's experience, it is also native to Australia where it has been an important food source for indigenous Australians throughout much of the continent, from Victoria to the Kimberley. The roots were cooked and the greens and stems eaten fresh, but perhaps of most value were the tiny seeds which were a staple, particularly in arid areas. Victorian colonist Robert Smyth wrote in 1878 of them being baked into cakes, 'infinitely superior to cakes made of nardoo flour'. Botanist Joseph Maiden noted in 1889 that, 'the food prepared from this seed must be highly nutritious, for … the natives get in splendid condition on it.' Purslane is another plant high in oxalic acid, so check our Note of Caution in Chapter 1.

Medicinally it has long been used for its proven abilities as an anti-inflammatory, an analgesic and a wound

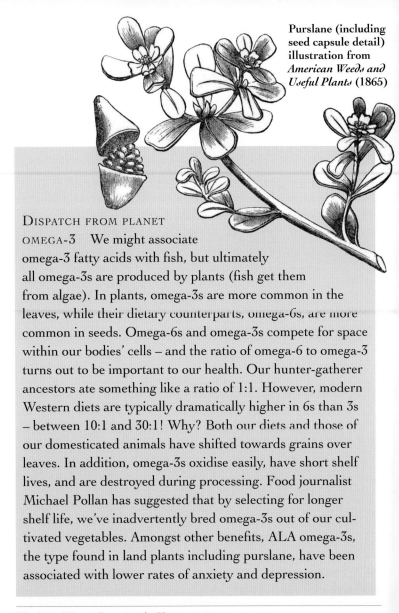

DISPATCH FROM PLANET OMEGA-3

We might associate omega-3 fatty acids with fish, but ultimately all omega-3s are produced by plants (fish get them from algae). In plants, omega-3s are more common in the leaves, while their dietary counterparts, omega-6s, are more common in seeds. Omega-6s and omega-3s compete for space within our bodies' cells – and the ratio of omega-6 to omega-3 turns out to be important to our health. Our hunter-gatherer ancestors ate something like a ratio of 1:1. However, modern Western diets are typically dramatically higher in 6s than 3s – between 10:1 and 30:1! Why? Both our diets and those of our domesticated animals have shifted towards grains over leaves. In addition, omega-3s oxidise easily, have short shelf lives, and are destroyed during processing. Food journalist Michael Pollan has suggested that by selecting for longer shelf life, we've inadvertently bred omega-3s out of our cultivated vegetables. Amongst other benefits, ALA omega-3s, the type found in land plants including purslane, have been associated with lower rates of anxiety and depression.

Purslane with and without flowers

healer, so it is great for bumps, sprains and scrapes of all kinds as a topical application. Herbalist Pat Collins calls it her 'summer chickweed' (purslane is available in the warmer months, chickweed in the cooler) for she uses it freshly crushed as a poultice on hot and itchy skin conditions.

LOOK FOR   Fingernail-sized, semi-succulent, mid-green oval leaves on plump stems, which are greenish bronze to start with, and then become quite red. The young plant has large leaves on stems that reach upwards (to 20 cm), but as growth continues the leaves produced are smaller and the plant tends to lie on the ground like a lacy mat. Tiny yellow flowers appear at the growing tips and in the fork between leaf and stem, and produce thousands of seeds like black grains of sand.

DISTRIBUTION   Purslane has an amazingly even distribution across the continent, from Tasmania to the dry centre to the Torres Strait Islands, with a tendency towards full sun and disturbed and bare earth. It seems to like growing in the cracks between loose paving, though seldom to a size worth picking from.

RELATIVES YOU MIGHT RECOGNISE   Purslane is in the family Portulacaceae, which also includes the larger, domesticated purslane sub-species golden purslane (*P. oleracia* subsp. *sativa*), the dwarf jade plant (*Portulacaria afra*), as well as several ornamental *Portulacas* found in nurseries.

Salsify
Tragopogon porrifolius

ALSO KNOWN AS Purple Salsify, Oyster Plant, Goatsbeard

A recently emerged salsify, a potential root harvest specimen

Salsify is also known as the oyster plant, as within the fleshy taproot of some plants there can be discerned a seafood-like flavour. We've tried it baked, fried and boiled, and it certainly is delicious, although we are yet to stumble across any oysterish specimens. Sixteenth-century herbalist Gerard declared it 'a

A flowering salsify plant, looking less grass-like and more branching due to maturity

most pleasant and wholsome meate, in delicate taste farre sur-
passing either Parsenep or Carrot'. In the name of vegetable
kingdom egalitarianism however, we will choose to describe
it as 'equal to any of the popular pointy roots, and a little like
the love-child of parsnip and artichoke hearts, yet milder'. The
whole plant is edible, with the young shoots being collected in
spring as a green vegetable. Also (though why anyone would
bother doing this, we don't know), the root latex can be used as
chewing gum. File that under things to keep you busy after the
collapse of civilisation.

The roots are harvested in winter or spring before flowering,
after which they become too fibrous. They snap quite easily

Salsify illustration from
Flora Batava (1822)

upon tugging, so gently does it – or use a trowel. They look like slightly disoriented, small white carrots. The harder the soil, the more likely they are to fork into confusion; it's in softer soils that you'll find more rewarding examples. Give them a scrub with a stiff brush, peel off all knobbly or hairy bits and trim the ends, and they are quickly brought to order, ready to cook. Both leaves and roots exude a milky white sap when cut, but this has none of the bitter flavour you may expect, though it does cause root pieces to discolour very quickly. Drop them into a bowl of water spiked with lemon juice until you're ready to cook them. Keep it simple by steaming the pieces then sautéing in a little butter and soy sauce, or parboiling then roasting with olive oil, bay leaves, coarse salt and some smashed cloves of garlic. Or follow tradition and make them into a gratin, croquettes, fritters, or a cream of salsify and mushroom soup. They can become simultaneously mushy and fibrous if overcooked, so aim for tender but not collapsing. You can also roast the root, like dandelion or chicory, to make a coffee substitute.

Salsify root is rich in inulin, a sweet soluble fibre that is suitable for diabetics. Inulin is also thought to improve gut flora populations and aid in the absorption of calcium and magnesium.

In 18th century Britain, salsify went through a period of popularity as a root vegetable, and is still cultivated on a small scale in Europe and Russia. Varieties with larger roots are available, such as the eccentrically named 'Mammoth Sandwich Island'. Gardeners sometimes also plant it for its hardiness and striking flowers.

LOOK FOR   Salsify initially forms a 'basal rosette' – its dull-green grass-like leaves grow from a central point on the ground – and it's at this stage of life when it is best harvested, but hardest to recognise. Later in spring and summer, the upright stems and the distinctive flowers appear, reaching up to 1 m in height. It

The seed head with its wind-riding mini parachutes

Diego Bonnetto with a wild salsify root

exudes white latex if damaged, and produces a 3-5 cm flower of muted purple, with unusual bracts (the spikes of the green cup that hold the flower) that are longer than the petals of the flower. The seed head is not unlike that of dandelion, only larger. One commentator, seeing salsify for the very first time (next to a train line in Poland), entertained the theory that they might be dandelions 'genetically-modified in some way by electromagnetic forces from the pantograph cables'.

DISTRIBUTION   A native of southern Europe and northern Africa, salsify can be found south of the tropics in all states of Australia except the Northern Territory. It prefers full sun and can handle dry conditions. Perhaps less common than others we've mentioned, but a special treat. Look out for them in grasslands, roadsides and vacant lots.

RELATIVES YOU MIGHT RECOGNISE   Spanish salsify (*Scorzonera hispanica*) is not in the same genus, but is a close relative all the same. It has a black-skinned taproot which is considered excellent eating.

Sow Thistle

Sonchus oleraceus

A good size for picking

Sow thistle's Latin species name *oleraceus* denotes a 'vegetable used in cooking', and this one certainly has been: from Malta to China to Maori societies, there are local names and recipes for this plant across time and geography. The leaves, which are rich in vitamins, iron and calcium, have an agreeable, slightly bitter taste which, though it intensifies in older plants, is lost when cooked. Tender leaves from young plants (and the yellow

petals) can be used raw in salads or cooked liked spinach. From mature plants pick the top 15 cm or so, including flower buds – if it snaps off cleanly it will be delicious sautéed with olive oil and lemon. Check for aphids, which adore plants of this genus. You can curse this trait, or let these living pest traps attract beneficial insects.

If you have never seen a sow thistle before, you may think us a little overly devoted to eating wild foods in this case, but *Sonchus* species are not true thistles and have no painful spikes. The 'sow' relates to the fondness pigs have for it. We grew up calling it 'milk thistle', a tribute to the white latex it exudes when broken. However, there is a true thistle, the powerful medicinal plant *Silybum marianum*, which is more properly called milk thistle, and is covered in Chapter 3, Other Weeds.

Medicinally, sow thistles possess antioxidant, anti-fever and anti-inflammatory properties, and have a long and wide history of varied medicinal use. A 13th century English herbalist recommended a diet of sow thistles 'to prolong the virility of gentlemen', no doubt widening the plant's popularity in the late Middle Ages. Pliny recommended it for bad breath, and famous 17th century herbalist Nicholas Culpeper promoted its milky sap for cosmetic uses, noting that it 'is wonderful good for women to wash their faces with, to clear the skin and give it lustre'. Based on the 'doctrine of signatures' (the belief that plants can be used to cure ailments of body parts they resemble), Culpeper wrote, 'The decoction of the leaves and stalks causeth abundance of milk in nurses, and their children to be well-coloured.'

Sow thistle illustration from *Deutschlands Flora in Abbildungen* (1796)

Many European explorers of Australia mention seeing and eating it, and leading ethnobotanist Beth Gott and representatives of the Yorta Yorta people consider it to be native – although that view remains contentious, as there are edible native relatives. At the very least, indigenous peoples were quick

DIY SUPERPOWERS In an ancient Greek precursor to Popeye's spinach popping, Theseus, the mythological founder of Athens, strengthened himself with a bowl of sow thistle before his encounter with the Bull of Marathon. E.C. Segar, the creator of Popeye, chose to promote spinach because of its high vitamin A content, not iron as is popularly believed – but either way sow thistle is higher in vitamin A and has double the iron, so Theseus chose the better vegetable.

Note to kids Eat your greens before fighting bulls (and remember your pants)

to adopt it as food: in the Yorta Yorta language it is known as *buckabun*, and in Woiwurrung it is *dalurp*. In 1878 Smyth reported the Kurnai mythology 'that the soul, as soon as it leaves the body, goes off to the east, where there is a land abounding in sowthistles (Thallak), [in] which the departed eat and live.'

LOOK FOR This plant is a shape shifter: it changes dramatically as it matures. And yet we have found that our Edible Weeds Walks participants rapidly pick up on the progression after seeing a few plants at different growth stages, and rarely

Sow thistle seedling – note the rounded leaves on the younger plants. This often correlates with better flavour.

have trouble identifying it successfully. Perhaps this is testimony to our many millennia of plant observation in the name of survival. When young, the thin tender leaves are a misty matt green and have a spade-shaped tip, followed by one or more smaller arrow shapes. As the hollow stem elongates and begins to branch, the leaves develop a blue undertone, become more jagged, coarser, and slightly shiny, and the plant looks more like an upwardly mobile dandelion. The yellow flowers too are dandelion-like, and also turn into pale fluffy seed heads, although they never form a sphere as with dandelions, and multiple flowers are born on the stalks instead of just one. By the time the seeds have developed, the plant can be up to 1m tall, and often has a purple tinge to the leaves.

DISTRIBUTION   A native to Eurasia, sow thistle is one of the most widespread weeds of the world. It occurs across wide-ranging conditions throughout most of Australia.

RELATIVES YOU MIGHT RECOGNISE   A close relative, prickly sow thistle (*Sonchus asper*) is also widespread and traditionally used interchangeably with common sow thistle for medicinal purposes. Prickly sow thistle is also edible, but a bit less delicate, with more thistle-like leaves. There are two native species that are also close relatives (*S. hydrophilus* and *Actites megalocarpa*), both used by aboriginal peoples as food.

Wild Brassicas
Brassica species

Some fresh, young, wild brassica greens

Weeds are the perfect candidates for domestication: their short generations and copious seed production make developing desirable characteristics relatively easy. Some of the most notable examples come from the cabbage or *Brassica* genus, several species of which were taken from the wild, and cultivated by ancient farmers. Over centuries these European and Asian agriculturists shaped them into root, seed, flower, and leaf crops in a fantastical array of peculiar forms – from Romanesco broccoli to kohlrabi. The ancient Greeks grew three

Brassica varieties, and Greek mythology reports that the cabbage first sprouted from where Zeus' sweat hit the ground. This may account for the slightly sulphuric aroma of cut cabbage, but they taste surprisingly good considering …

Wild brassica varieties still abound: there are at least ten species in Australia. Broccoli, cauliflower, kale, Brussels sprouts and cabbage are all forms of *Brassica*. The wild relatives closest to these in flavour are *Brassica fruticulosa*, and wild forms of *B. rapa* and *B. oleracea*. The young plants look almost indistinguishable from broccoli seedlings, and the taste is pleasant and immediately familiar.

Other wild brassica species found in Australia include the liberated siblings of the cultivated Indian mustard (*B. ×juncea*) and black mustard (*B. nigra*). The fiery flavour of their seeds can also be found in their leaves. In Nepal and the Punjab region of India and Pakistan, mustard greens rather than spinach are the traditional ingredient in saag curries. Other species, such as the widespread wild turnip (*B. tournefortii*) may be less palatable – a little tougher and hairier – but all *Brassica* species, and all parts of the plants are considered edible. Wild brassicas come in all textures, shapes and colours (well, green to purplish green) but since none are toxic, you can go forth, taste and experiment – once you know the basic identifying features.

We have a soft spot for the yellow flowers, which are like teeny broccoli heads you can nibble on while foraging, or add to a salad. Unless very soft and young, the leaves are generally best cooked – many develop a lovely velvety texture. Cook into Asian noodle soups, pan-fry with potato, onions and haloumi, use wilted with poached eggs and tomato relish – how you use this leaf will vary according to variety, but if you think of a

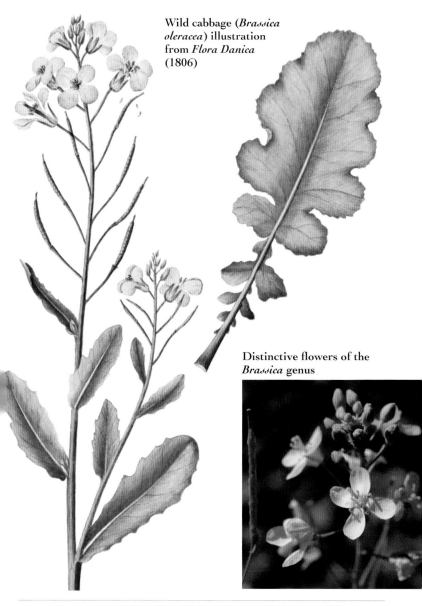

Wild cabbage (*Brassica oleracea*) illustration from *Flora Danica* (1806)

Distinctive flowers of the *Brassica* genus

'cabbage to mustard greens' flavour spectrum, the classic culinary accompaniments will do you nicely.

There's no nutritional data available specifically for the wild brassica varieties, but if we were to use kale as a stand-in, then within the confines of this book, and perhaps the confines of the world, wild cabbage is third only to dandelion and parsley in terms of its nutritional density. Brassicas are a good source of many health-promoting substances including folic acid, phenolics, carotenoids, selenium and vitamin C.

Medicinally, the *Brassica* tribe is perhaps most famed for its anti-cancer properties. Regular consumption is associated with reduced lung, colon, prostate and other cancers. Since wild plants tend to be higher in antioxidants, the wild forms may be more potent. Brassicas may also assist in preventing and ameliorating the symptoms of diabetes.

In the garden, mustardy brassicas are an excellent natural pest-control agent. Chop the plants roughly with a shovel and dig into the soil in spring as a 'bio-fumigant' for up to 40% more produce from your *Solanaceae* crops (e.g. potatoes, tomatoes and eggplants), particularly if you have grown these in the same spot the year before. The technique is even used on commercial farms, offering a CSIRO-recommended alternative to banned synthetic pesticides such as methyl bromide.

LOOK FOR   Wild brassicas range in shape, but are united by their four-petalled lemony yellow flowers growing along the tips of the flowering stalk. They have distinctive elongated seed pods (1-2 cm long) that contain a single line of tiny seed, much like miniature peas in a pod. Most have ripple-edged leaves with a blue-grey tinge often becoming purplish in old age. In general choose those with tender silky leaves, rather than the slightly hairy dimpled ones.

A particularly lush, mature wild brassica

In practice there are many edible look-alike relatives in the broader Brassicaceae family that are difficult to tell from true brassicas. Fortunately none (to the limit of our knowledge) are toxic and many are quite delicious. Those of us in the south get to experience the wonderfully spicy wild rocket (*Diplotaxis tenuifolia*).

DISTRIBUTION   Collectively, the wild brassicas have a wide distribution across all states and climates, but are less common in the tropics. They are adapted to disturbed soil of all types, and grow in full sun.

RELATIVES YOU MIGHT RECOGNISE   Other famed brassicas we haven't mentioned include turnips, wombok and bok choi (all *B. rapa*), and the oil crop canola (*B. napus*).

Wild Lettuce

Lactuca serriola

Also known as Prickly Lettuce, Opium Lettuce

Wild lettuce is a lovely delicate green, and ancestor of many forms of cultivated lettuce. Its taste is indistinguishable from that of its descendants and it can be used in just the same ways. Being more substantial than many lettuces, it is also suitable for cooking. You must pick it young while the leaves are tender, crunchy and mild. At this stage it grows as a 'rosette', the leaves emerging from a central point at ground level. It can be found in this form generally between autumn and early spring. Past this appetising stage, the plant undergoes

A young wild lettuce, good for eating

An older plant, way past good-eating time!

Wild lettuce illustration from *Deutschlands Flora in Abbildungen* (1796)

Wild lettuce seeds

quite a Jekyll-and-Hyde transformation: it grows an upright prickly central stalk, and the leaves become leathery, intensely bitter and spiny!

When cut, the plant weeps a white latex known as lactucarium. This has analgesic and sedative qualities, for which it gained the name 'lettuce opium' (despite containing no true opiates). Hippocrates described these mild narcotic effects in 430 BC. Pliny, in the 2nd century, wrote of the lettuce's ability to suppress sexual desire. Yet, paradoxically, the ancient Egyptians used wild lettuce for just the reverse: as an aphrodisiac. Italian ethnobotanist Giorgio Samorini believes he has found the answer: 'Tests showed that 1 gram of lactucarius induces calming and painkilling effects because of the presence of lactucin and lactucopicrin … At the highest doses [2 to 3 grams], the stimulating effects of tropane alkaloids prevail. This finally solves an ethno botanical riddle.' Way to go science!

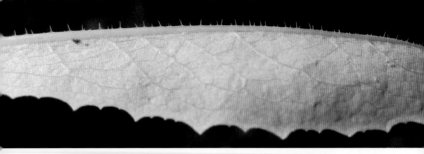

Look for the distinctive row of hairs along the back of the spine of a young leaf. These become even more pronounced on the bitter leaves of older plants.

LOOK FOR   When young, the mildly serrated leaves grow from a central point at soil level, and the plant looks a little like a brighter dandelion. Each leaf has a distinctive row of spikes (not at all prickly at this stage) running along the spine on the leaf's underside. As it ages, the plant forms a central stem, the leaves stiffen and take on a darker and more bluish tinge, the spikes become hard and prickly, and small yellow flowers are produced. The whole plant exudes white latex when broken at any age.

DISTRIBUTION   Native to Europe and Asia, it is found in all the states of Australia, from Hobart to the tropics. It grows widely, in conditions ranging from desert to garden and shady riverbank. The tastiest plants grow in moist, fertile soils and often in partial shade.

RELATIVES YOU MIGHT RECOGNISE   The genus *Lactuca* also includes all garden lettuces, most of which are varieties of *Lactuca sativa*.

3 | OTHER WEEDS

Plants included in this chapter have definite merits but are generally less common, harder to harvest or process, or offer some other impediment to free and easy enjoyment by comparison with those in Chapter 2. Some are medicinal-only or have poisonous parts, so don't go eating just any bits that take your fancy. This short selection far from covers all this wide country's other weedy edibles and medicinals. It is merely a taster.

Briar Rose
(*Rosa rubiginosa*)

Both the petals and the hips of this wild rose can be used in cooking: jellies, jams and syrups are favourites. Briar rose hips are tastier than the fruit of most cultivated roses, and home-made rosehip tea is as simple as simmering 7-10 hips in water for 5 minutes. The hips should be collected in summer and autumn when they are deep red and soft. Use the whole hip if you are simply extracting its 'vital essences', as for syrup, jelly or tea. But, if you are making jam or using the actual flesh in other ways, you will have to split the hips and scrape out the seeds and hairy fibres that surround them, as these are an intestinal irritant. The pink flowers of this prickly shrub are modest by cultivated-rose standards, with a single row of petals.

They are found in the southern states, extending into southern Queensland, as a weed of roadsides, grazing land and disturbed bushland.

Calendula or Pot Marigold
(*Calendula officinalis*) and
Field Marigold
(*C. arvensis*)

Commonly planted as vegetable garden companions, both the leaves and flowers of these unfussy plants are edible. We prefer the very young leaves, cooked. The petals have a slightly peppery sweetness and, apart from being great raw as a salad ingredient, they can be used in fritters, broths or creamy dishes to add flavour and impart a gorgeous golden colour. Feed them to your chooks for egg yolks so yellow you'll need sunglasses to dunk your toast soldiers in them. Famous for anti-inflammatory and wound healing properties, they are often made into salves

and ointments for use on burns, cuts and chapped skin. Calendula is mostly a garden escapee, while field marigold is a more widespread weed of fields and wastelands.

Chicory
(Cichorium intybus)

This bitter green can be used cooked and in salads in much the same way as its cousin, the dandelion. The roasted roots have been used as a coffee-style drink for centuries. Chicory has a

variety of uses in traditional medicine, including as a tonic for the liver. In Australia it is both a garden plant and a widespread weed, especially of fields and roadsides south of the tropics. It can be recognised by its distinctive blue flowers.

Cobbler's Pegs
(Bidens pilosa)

Cobbler's pegs, also known by a range of curse words, is notorious for its seeds, which stick to trousers and shoelaces. This scrambling South American native is one of the most important wild greens in eastern Africa. You can add the tips and young greens to soups. The resinous taste is not to everyone's liking (only one of us enjoys it). They are high in antioxidants and have many uses in folk medicine including the use of the juice for gastric ulcers and as a wound dressing. Cobbler's pegs is widespread, but less so in the south of the country. Look for the dis-

tinctive black 'sea urchin' seed clusters, as the small daisy flowers can have white, yellow, or even absent petals. One of its look-alikes is another tasty weed: gallant soldier (*Galinsoga parviflora*).

Docks and Sorrels
(*Rumex* species)

Plants from the large Rumex genus are referred to as either docks or sorrels. They are all edible, but with flavours ranging from scrumptious to unbelievably bitter. When good, these plants are a delicious, strongly lemony cooking green. Pinch off a corner to test taste before you bother picking. Sheep sorrel (*Rumex acetosella*), with its small lance-shaped leaves, is the most consistently good. It often grows on acid soils and in pasture. Its larger cousins, the various docks, are particularly prevalent near waterways and on heavier soils, and vaguely resemble a lighter-green silver beet. All are high in oxalic acid (see our Notes of Caution in Chapter 1).

Dock

Sheep sorrel

125

Fleabane
(Conyza bonariensis, Conyza canadensis and related species)

We use fleabane as the name prescribes: to remove fleas from our dog. Boil large bunches in water, and add the strained-off liquid to the bath when washing your canine. Many human medicinal qualities have been attributed to it too, most widely as a tea to treat diarrhoea. Fleabanes begin as low rosettes that shoot skywards with a single stalk, and produce lots of tiny pom-pom seed heads. A very widespread weed, it is often the first to sprout from cracks in the concrete.

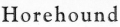

Horehound
(Marrubium vulgare)

Horehound has long been used as a juice or in syrups and teas as a remedy for coughs and chesty colds. Horehound contains a substance called marrubiin, which is an anti-spasmodic analgesic believed to loosen bronchial phlegm. It may stimulate the uterus so don't take it during pregnancy. Looking much like its cousins the mints, horehound is a widespread weed, often of overgrazed country.

Milk Thistle
(Silybum marianum)

This is an intimidatingly spiky thistle with white markings, as if milk has been spilt on it. It is considered a liver tonic, and is commercially available as such in capsule form. It is also the most powerful known cure for eating the death-cap mushroom (*Amanita phalloides*) – which otherwise means death or a liver transplant. Milk thistle is widespread, but more common in the countryside than the city. The name milk thistle is also applied to sow thistle (*Sonchus* species, see Chapter 2).

Petty Spurge
(*Euphorbia peplus*)

Not to be eaten, this diminutive and fragile looking weed produces a fiercely skin-burning white latex when cut. An extract of petty spurge has been developed commercially for treating actinic keratosis (the flaky pre-cancerous 'sun spots'). However, there are more effective medications. It can be used to treat warts though. The home user can dab a drop of the sap onto the affected spot once or twice a day, washing off any that

ends up where it shouldn't. Be *very* careful not to get any near your eyes. It reputedly works on warts too. Petty spurge is very common and found in most gardens, anywhere from southern Queensland down.

Shepherd's Purse
(Capsella bursa-pastoris)

The little heart-shaped seed heads of shepherd's purse have a peppery-mustardy taste, while the edible leaves and flowers are milder. All are good in salads. It is widespread, and often grows in parks and ovals, where mowing means this small plant is kept even smaller, making the seed heads the best fare. It is considered good for stopping both internal and external bleeding – on one of our walks, an elderly German woman remembered collecting it for soldiers during WW II. A tea is used to stop uterine bleeding.

Stork's Bill
(*Erodium cicutarium*)

The young leaves of this geranium relative taste a little like carrot tops. They are best chopped finely and cooked. The close relative musky stork's bill (*E. moschatum*) is equally common, difficult to distinguish but also edible and, as the name suggests, has a slightly musky aroma. These plants can be prolific in pastures and expanses of mowed grass, and grow widely south of the tropics.

Sweet Violet
(Viola odorata)

The flowers of all violets are considered edible, and are used crystallised in cake decorations or made into syrups. In cities, the most common is sweet violet, which likes damp, fertile, shady places, and can be found in all the capitals except

Darwin. It has mild leaves, which can be eaten raw when very young; otherwise they can be used cooked, and have a slightly thickening effect in soups, similar to mallow. Flowers can be added to salads or frozen into ice cubes for summer drinks. Some sources mention the seeds and rhizomes of violets causing gastric discomfort and other symptoms, so avoid these parts.

Wild Celery

(Apium graveolens)

Wild celery is the same species as cultivated celery. Some-times, weedy versions are recent garden escapees, while others have spent longer re-wilding, and have thinner stalks and a potently delicious flavour. Chop stems and leaves finely for a memorable salad experience or for use in soup stocks. Wild celery is usually found near creeks and streams, so avoid it if pollution is an issue. It is a minor weed of urban areas, and is not found in the tropics.

4 | WEEDY RECIPES

These recipes aim to hit the mark on several levels: they are healthy, delicious, easy and economical, much like the weeds they revolve around. We have also tried to work largely with foods that many people will already have in their pantries – a practical homage to the foraging concept of using and enjoying what is right in front of you rather than buying specialty products to fit a whim.

We firmly believe recipes are made to be disregarded, and encourage you to get a feel for the weeds' individual charms and cook with them as your palate dictates. There are suggestions for use (traditional and our own) dotted through the profile of each weed to provide inspiration should you crave it.

Green Smoothie

This may be the healthiest thing you can eat. Ever. Call it
Monster Powers Juice and it is a brilliant way to get kids to
eat masses of greens. For the slightly more mature amongst us
it also seems to do quite a nice job on a mild hangover. Mix it
up by adding grated ginger or chopped mint.

INGREDIENTS FOR 2 SERVES

1 medium banana (frozen is especially nice)
1 medium orange or half a mango – or use a whole mango and
    skip the banana
1 cup of water
½ to 1 cup, depending on taste, of mixed weeds: chickweed,
    cleavers, dandelion (keep this one to a minimum if you dislike
    bitter flavours), mallow, nettle, plantain and sow thistle all
    work well.

Avoid fennel, nasturtium
and more than a skerrick
of purslane in this recipe,
as their strong flavours will
really take over.

METHOD

Wash and de-stem all the
weeds, and chop across the
grain. Peel and slice the
orange, removing any pips,
and break the banana into
chunks. Place all ingredients
in a blender and whizz until
smooth.

Purslane Yoghurt Dip

This is essentially a variation on tzatziki, and in all earnestness we declare it a superior one. If you plan on kissing someone who hasn't shared this with you, you may want to add half a cup of very finely chopped flat-leaf parsley to offset the effects of the garlic (also add a splosh of extra yoghurt and olive oil in this case). Otherwise, experiment with adding a tablespoon of very finely chopped mint, dill or wild fennel.

INGREDIENTS FOR 4 SERVES
1 tightly packed cup of washed purslane, stems mostly removed
½ cup good quality, plain yoghurt
2 cloves garlic, minced
¼ teaspoon salt
1 tablespoon extra-virgin olive oil
Turkish bread

METHOD
In a bowl, mix together the yoghurt, oil, garlic and salt, then fold through the chopped purslane. Put in the refrigerator to chill. Serve with hunks of fresh bread.

Mixed Weeds Salad

This is pure, (nearly) unadulterated weedy goodness. The nutty crunch of seeds plus the sweetness of the sultanas really help to set off the flavours of the wild greens, and even the cynical polish this off with lip-smacking fervour during our edible weeds courses. It looks very nice too.

Chickweed, wild lettuce and nasturtiums are wonderful when in season. For amaranth, dandelion, fat hen, mallow, plantain, sow thistle and wild brassica, pick the youngest, most tender leaves for salad eating. You can also add purslane and a little angled onion or fennel, both chopped very finely so as not to dominate.

INGREDIENTS FOR 4 SERVES AS A SIDE DISH
3 tablespoons mixed sunflower and pumpkin seeds (*pepitas*)
2 tablespoons sultanas
6 teaspoons tamari
4 tablespoons extra-virgin olive oil
2 tablespoons balsamic vinegar
2 tablespoons lemon juice
salt and pepper to taste
mixed weeds, enough to fill a large salad bowl
petals of 3 calendula flowers (or equivalent other edible flowers
   – wild brassica or nasturtium flowers are delicious)

METHOD
Toast the sunflower and pumpkin seeds with the sultanas in a dry pan over a low heat. As they begin to colour, sprinkle with two teaspoons of tamari, and keep stirring gently until they are coated and the liquid has evaporated. Allow to cool completely.

Mix the olive oil, remaining tamari and lemon juice in a small jar and shake well. Add salt and pepper if desired.

Wash and de-stem all weeds, pat dry thoroughly, and chop across the grain. Toss in the bowl with the dressing. Sprinkle the toasted seeds and sultanas and the calendula petals over the top immediately before serving.

Weedy Frittata

Frittatas were surely invented as a glorious way to use up a glut of eggs by wrapping them around a glut from the garden or some tasty leftovers. Add sliced, steamed potatoes or diced roasted vegetables to make this even more of a meal. For glamour, decorate the surface with thin circles of tomato or strips of roasted capsicum just before popping it under the griller.

Fat hen in particular and nettle (see under Nettle in Chapter 2 for tips on handling) go exceedingly well with eggs. The bitter note of dandelion is lovely here, and amaranth, mallow, plantain, sow thistle and wild brassica are all good inclusions too.

INGREDIENTS FOR 4 SERVES
1 tablespoon olive oil
6 eggs
¼ cup milk
¼ cup grated parmesan
1 teaspoon rosemary, very finely chopped
1 teaspoon sweet paprika
salt and pepper to taste
1 cup angled onion (Don't use the top two-thirds of the plant, as this becomes fibrous with cooking. If not in season, substitute with 1 clove of garlic plus onion or leek.)
4 cups mixed weeds
½ cup grated cheddar

METHOD
Keeping angled onion separate, de-stem and wash all weeds, and chop across the grain. Steam all but angled onion until just tender, then squeeze out all excess liquid.

Whisk the eggs, milk, parmesan, rosemary, and salt and pepper together in a bowl.

Heat the oil in a heavy-bottomed frying pan, and fry the angled onion with a pinch of salt until softened. Stir the weeds evenly through the egg mixture, then pour it over the angled onion in the frying pan. Swirl paprika through the mixture with a fork. Reduce to a low heat and cook for 5 minutes, or until mostly set.

Sprinkle cheddar over the top, and put the pan under a griller on low heat. Cook until the cheese has melted and turned golden – takes about 3 minutes. (*Tip* If your frying pan has a wooden or plastic handle, wrap a little foil around it where it will be exposed to the griller.)

Nettle Gnocchi

These little dumplings are a nettle classic. Walnuts pieces are often added to the sauce in this sage butter version, while a mixed herb cream sauce with a splash of brandy is also excellent.

INGREDIENTS FOR 4 SERVES

THE GNOCCHI

½ kg floury potatoes (King Edward, sebago, coliban)

150 g nettle leaves (about 3 cups, fairly firmly packed)

⅔ cup grated parmesan

2 free-range egg yolks

1-1½ cups of plain flour, plus extra for rolling

salt and pepper to taste

THE SAUCE

2 tablespoons olive oil

2 tablespoons butter

3 tablespoons chopped fresh sage leaves, or 2 of mixed thyme and rosemary

3 cloves of garlic, finely chopped

METHOD

THE GNOCCHI   Peel and chop the potatoes, steam until tender, then mash until smooth.

Drop the nettles into a pot of boiling water for 2 minutes, then transfer them to a bowl of iced water for another 2 minutes. Drain and squeeze out the excess water with your hand. Chop very finely, then stir into the potato mash.

Add half the cheese and both the egg yolks, and season well with salt and freshly ground black pepper. Add 1 cup of flour and quickly work it into the potato. When the dough feels not too sticky to roll out, cut off a small piece, roll it into a ball and drop it into boiling water to test. If it floats to the surface and holds its shape well, your dough is ready. Otherwise incorporate a little more flour and test again.

Generously flour your work surface and roll the dough into long sausages about the thickness of your finger, and cut off pieces 2 cm long. As you go, keep the gnocchi on a tray with a little flour sprinkled over them.

To cook, drop batches into a large saucepan of boiling water. Be careful not to overcrowd the pot, or the gnocchi will stick to each other. They are done when they float to the surface (generally within a minute). Remove with a slotted spoon and tumble them in a little olive oil to prevent sticking.

THE SAUCE   In a pan, melt the butter with olive oil, and fry the sage and garlic until just crispy, then stir in the gnocchi to warm them through.

Sprinkle with the remaining cheese to serve.

Egg and Nasturtium Sandwiches

A twist on that classic member of the Sandwich Hall Of Fame, this is just as zesty and definitely more colourful. You could use conventional capers, but using nasturtium 'capers' is more fun, and makes the sandwich like a theme party for nasturtiums.

Nasturtium capers are simply the plant's pickled flower buds and seeds. Choose totally unopened buds, and seeds that are still fresh and green. Measure your haul of seeds and buds in a cup, and bring the same volume of white wine vinegar to the boil with a good pinch of salt and a few smashed peppercorns; then pour this hot mix into a jar over the 'capers'. You can add a bay leaf, dill, or tarragon if you like. Seal and leave for at least a month.

INGREDIENTS FOR 1 SERVE

2 hard-boiled eggs

2 tablespoons mayonnaise

½ cup nasturtium leaves and
 flowers (in equal parts)

1 tablespoon chopped angled onion

1 tablespoon chopped nasturtium
 'capers' (for recipe, see above)

salt and pepper to taste

2 large slices of very good bread

METHOD

Chop the eggs and nasturtium leaves and flowers finely. Combine them in a bowl with all the other ingredients (except the bread!), and mash with a fork until well mixed. Apply to bread.

Moroccan Mallow Stew

Mallow is a much-loved wild green in the Middle East, and is frequently paired with coriander, garlic and lemon. *Molokhia* (a thick mallow soup served with rice and chicken or lamb), and *bakoula* (a mallow purée with red olives and preserved lemon, used to dunk bread in) are both classic dishes.

INGREDIENTS FOR 4 SERVES

5 cups tightly packed young mallow leaves, de-stemmed and washed
⅔ tin or 1 cup cooked chickpeas, drained and rinsed
1 tin or 1½ cups cooked chopped tomatoes
zest of 1 lemon, plus the juice
3 cloves garlic, minced
1 teaspoon salt
3 tablespoons chopped dates or sultanas
2 teaspoons each ground coriander and cumin
large pinch of chilli flakes
2 tablespoons extra-virgin olive oil
1½ cups couscous
½ cup good quality plain yoghurt
4 tablespoons flaked almonds

METHOD

Roughly chop the mallow leaves, and place all ingredients (except the oil, almonds and couscous) in a deep pan with a heavy base. Simmer covered for 10 minutes, then uncover, stir and simmer for a further 2-3 minutes. Stir the oil through immediately before serving.

While stew is cooking, toast the almonds in a dry pan until pale brown.

Boil 2 cups of water with a big pinch of salt. Take off the boil, pour in the couscous, stir, and cover with a lid for a few minutes. Remove the lid and fluff with a fork.

Place mallow mix on a bed of couscous. Add a dollop of yoghurt and sprinkle with toasted almonds to serve.

Wild Cabbage and Eggplant Stir-fry

This is an unabashedly pan-Asian, unabashedly scrumptious vehicle for the tangy, mustardy medley that is oxalis and wild brassica. That said, you could use any of the cooking greens mentioned in this book instead, but we think this one has the most kick for offsetting the creamy mildness of the eggplant and tofu.

INGREDIENTS FOR 2 SERVES

1 tablespoon sesame seeds
1 small eggplant
4 cups firmly packed wild brassica greens, washed, and chopped roughly across the grain
1 cup firmly packed oxalis leaves, washed
a small knob of ginger
2 cloves of garlic
1 small red chilli
1 tablespoon vegetable oil
2 teaspoons sugar dissolved in 3 tablespoons tamari
1 teaspoon sesame oil
200 g firm tofu, cut into 5-mm thick slices, or 2 small fillets of mild white fish
¾ cup jasmine rice (about 2 ½ cups when cooked)

METHOD

Toast the sesame seeds in a dry pan on a low heat, and set aside. Start cooking the rice.

Slice the eggplant to finger thickness, and steam until tender. Add the wild brassica to the steamer for the final minute, then remove to its own bowl after steaming.

Finely chop the ginger, garlic and chilli. Heat the vegetable oil in a frying pan, and fry aromatics and eggplant slices until they brown a little.

In a separate pan, lightly fry the tofu or fish pieces in a little oil with 1 tablespoon of the tamari and sugar mixture.

Add the brassica and oxalis with the remaining tamari mix to the eggplant pan and toss all together on the heat until the greens are wilted. Remove from heat and fold in the sesame oil.

Place the vegetable mixture on a bed of rice, top with tofu or fish, and sprinkle with toasted sesame seeds to serve.

Prickly Pear Pizza

We love this. We made it up, then we wanted to make it all the time. We made it for other people, then *they* wanted us to make it all the time.

It is worth cooking up several cactus pads at once in the way suggested here, as they keep well in the fridge and can then be thrown into other dishes throughout the week.

Tip To simplify this recipe, use Lebanese flat bread as a pizza base and halve the time the pizza spends in the oven.

INGREDIENTS FOR 4 SERVES

THE TOPPING
½ cup of your favourite pizza sauce
2 cloves garlic, minced
1 teaspoon mild chillies, very finely chopped
1 medium prickly pear pad
150 g feta cheese
⅔ tin *OR* 1 cup cooked kidney beans, drained and rinsed
salt and pepper to taste
½ a lemon for squeezing

THE BASE
1½ cups plain flour, plus extra for kneading and rolling
1 teaspoon (½ a 7-g sachet) dried yeast
½ teaspoon salt
¾ cup lukewarm water
2 teaspoons olive oil, plus extra for brushing

METHOD

THE TOPPING   De-spine your cactus pad (for tips, see under Prickly Pear in Chapter 2) and slice it into pieces about 2 x 1 cm. Boil in well-salted water for 7-8 minutes, then drain. Soak the

kidney beans overnight then cook in salted water until tender, or rinse well if using tinned beans.

THE BASE   Thoroughly combine flour, yeast and salt in a large bowl, then make a well in the centre of the mix and add luke-warm water and olive oil.

Use a wooden spoon to mix, then use your hands to bring the dough together in the bowl. It should be soft and just a little sticky. Turn onto a lightly floured surface and knead until smooth, (about 3 minutes). Don't over-knead.

Return the dough to the bowl and lightly coat with olive oil to prevent drying. Cover the bowl with a damp tea towel and place in a warm spot to rise. Leave for about 1 hour, or until doubled in size.

Brush two medium pizza trays with olive oil and sprinkle with a little flour. Preheat the oven to 230°C. Divide the dough into 2 equal portions and roll out on a lightly floured surface to almost the size of your pizza trays. Lift onto trays and tease the dough out to the edges with your fingers.

Prick with a fork, avoiding the edges, and apply your pizza sauce followed by the beans, then prickly pear, then crumbled fetta. Scatter chilli and garlic across both pizzas, season with salt and pepper and finish with a squeeze of lemon juice.

Leave in a warm place for 15 minutes or until the edges have risen slightly. Bake in the oven for 10 minutes, then swap the trays around in the oven and bake for a further 5-10 minutes.

5 | WEEDS IN THE GARDEN

The plants we call weeds are an irrepressible force of nature. Absolute control – that is, total eradication – is very often not an option, and we can waste a lot of energy or put a lot of money into the hands of chemical companies trying. Read on, and discover a more fruitful approach …

Weeding is one of the perennial chores of the gardener. By discovering and utilising the merits of these spontaneous guests, we immediately give ourselves more time to stare absent-mindedly into the middle distance. Weeds' merits do not end with their edible and medicinal qualities. There are several ways they can benefit our gardens too, but to manage them to our gain and with minimal effort, we need to first understand a little of their dispositions.

The key phrase here is 'nature abhors a vacuum'. If we work from the starting point that ecologies are in a constant state of attempting to fill every available niche, we begin to understand both why weeds appear in our gardens and how they can be useful to us.

While gardening we often strip and dig over earth. Yet naked fertile soil is an aberration in the natural world, and life will attempt to place an ecological 'band-aid' over that ground from the moment we expose it. The fastest growing plant band-aids tend to be weedy species, whose abundant seed banks can help put a green fuzz on barren earth. And that immediately begins

Colonising bare ground

the processes of retaining moisture, creating organic matter, improving soil structure, and feeding soil microbes. If you don't want weeds to move in to do these essential jobs, you must cover the soil in some other way. But why not let them do what they do so well, and exploit them a little while they are at it?

WEEDS AS PLANTS TO IMPROVE THE SOIL

Picture this: you have a garden with lots of bare earth, and hard compacted soil. The soil has little structure, organic matter, or micro-organism life to keep it healthy, and most nutrients have been leached out of the topsoil into deep subsoil. Rain runs straight off, and any moisture that is retained evaporates in no time. Much of what you've planted seems stunted and unhealthy.

Many weeds have vigorous root systems. Fibrous rooting weeds such as wild lettuce, brassicas and nettle can help stabilise sloping ground. Tap-rooted weeds such as dock, dandelion, plantain and mallow can penetrate hard or heavy soils, acting as nature's garden forks. Not only this, but all these roots remain in the soil and eventually decay, becoming that ultimate soil improver, humus. Decomposing roots also leave channels for drainage, aeration, and earthworms. Soil texture is improved, and those beneficial soil-inhabiting micro-organisms that are so essential for healthy plant growth can multiply. Above the ground, the decomposing parts of the weeds also contribute organic matter, which begins to form a mulch layer on the soil surface. In the case of deep-rooted weeds, they have accessed the nutrients from lower in the soil profile, and now release

The taproots of dandelion are harvested for roasting: one of nature's garden forks.

them into the topsoil as they decompose. The living weeds and this mulch layer act together to both improve the moisture retention and moderate the temperature of the soil – further helping those wondrous micro-organisms to thrive.

WEEDS AS NURSE PLANTS
All living plants will moderate the effects of wind and sun, and humidify and cool the air around young seedlings in hot, dry, or exposed gardens. As weeds will grow quickly and in adverse conditions, they are excellent casual labour in the role of shelter provision, and can be removed heartlessly once they have served their purpose! As your plant advances, cut back or remove the surrounding weeds and be sure to regularly chop their tops off to prevent them from going to seed. Inevitably, there is competition from the weeds for water and nutrients, but we have found that the benefits of this technique outweigh any losses.

GETTING RID OF WEEDS
No matter how impassioned you may become about the various weeds in your garden, you will inevitably want to get rid of some of them, much as you would want to get rid of your favourite armchair if it kept showing up in the middle of your bed. There are many excellent organic gardening books that can tell you how to do this, so we will only touch briefly on the subject.

Firstly, remove the conditions that promote weed growth by maintaining good levels of plant cover or mulch. Much of our garden – front and back – is devoted to what is known as an urban-scale 'food forest': strawberries, violets, nasturtiums and alyssums make a dense and colourful groundcover. French sorrel, rhubarb, asparagus, tansy, yarrow, blueberries, currants,

A 'food forest' allows little room for weeds to grow through.

acacias, and ultimately fruit trees rise above them. The system is so stacked and the soil so shaded that there are few niches left for any weeds to sprout.

Secondly, build a healthy, loosely structured soil. This means that weeds that do appear are much easier to pull out, and as

Some vegetatively reproducing plants, from left to right: couch grass, tradescantia, oxalis

healthy soil is the foundation of good gardening, your efforts will pay dividends on multiple levels.

We have found hand removal for incidental weeds, and 'sheet mulching' for bigger areas, to be effective for most weeds in the home garden. Hand removal might be pulling the weed up, or repeatedly chopping it off at ground level, thereby not disturbing the soil but gradually weakening the weed by preventing photosynthesis. These methods are not always enough for weeds that can reproduce 'vegetatively', that is, via root or stem pieces (runner grasses such as couch or kikuyu, and rhizome fragments of Jerusalem artichokes) or bulbs (such as oxalis). Oxalis can have vegetable oil drizzled very precisely (a syringe works well) into the soil at the plant's centre, which will coat the bulb and suffocate it. Runner grasses are best sheet mulched, but much more thickly than with other weeds.

An all-ages sheet mulching event on a permablitz

Sheet mulching involves feeding the weedy area (chicken manure is good) and watering it well. The idea here is to stimulate growth that moves some of the plants' energy from the roots into the leaves, which you are now going to suffocate. Lay wet newspaper at least eight sheets thick over the whole area, making sure the sheets have lots of overlap at every point. Then cover the newspaper either with a generous layer of mulch or with your new garden bed.

When you have pulled your weeds, you have a decision to make about their disposal to prevent them from reproducing: compost, 'weed tea' or the rubbish bin. If the weeds haven't gone to seed and aren't vegetative reproducers, then they are simply good nutrient-rich organic matter and can be used in any way you like. Composting in this case could mean a classic compost pile, or simply be a matter of chopping the weeds up, sprinkling them on an area that needs feeding, and covering them with some dry, brown material (woodchips, straw, dead leaves, branches) to prevent their goodness being lost to the air.

If your weeds *have* gone to seed or *are* vegetative reproducers, you need to deal with them in a different way. Only a hot-composting process will destroy these, and this involves building a fresh, layered heap of at least a cubic metre, so that its internal temperature can reach 55°C. Even then it is advisable to soak vegetative reproducers in a barrel of water for a few weeks first, stirring once a day with a stick to prevent them from becoming rank. When the liquid has become deeply coloured, strain it off, and you have made 'weed tea'. Dilute this until it has the colour intensity of apple juice, and use it to water plants and to spray on foliage for improved growth and health.

If this all seems too much effort, you could throw your weeds in the council green bin, but be aware that in doing so you are

A weed 'tea' – turn the problem weeds into a liquid fertiliser

stripping the soil of the nutrients that have gone into that weed – this can exacerbate soil deficiencies, and actually encourage a return of the same weed species that was so well adapted to deal with those deficiencies in the first place.

Further Information

We have used hundreds of sources, most of which are peer reviewed journal articles, medical books and archival texts. The full list can be found on our website www.eatthatweed.com, where you can also find additional photos, updates, walks and references.

The following selection of books and web resources have been particularly useful or inspiring, both over the years and in putting together this book.

CULTURAL USE OF PLANTS

BOOKS

Medicinal Plants in Folk Tradition: an Ethnobotany of Britain and Ireland, Allen & Hatfield, Timber Press, 2004

This is an entertaining source of folklore from Britain and Ireland.

Koorie Plants, Koorie People, Nelly Zola and Beth Gott, Koorie Heritage Trust, 1992

This, and other works by Beth Gott, have provided us with information about local indigenous usages.

WEBSITES

Internet Archive and Open Library are two vast digital libraries containing out-of-copyright texts, from which we've gleaned much historical information: www.archive.org and www.openlibrary.org

Google Scholar is a subsection of the search engine behemoth focused on scientific and other academic journal articles. A search for 'ethnobotany' along with a plant name is a good way to research traditional uses from around the world, although unfortunately you will need a university log-in to access many of the articles: scholar.google.com

BOOKS

Edible Wild Herbs of Australia and New Zealand, Tim Low, Angus & Robertson, 1991
  Out of print so, sadly, hard to get, but a very thorough book if you can find it.

Edible Wild Plants: Wild Foods from Dirt to Plate, John Kallas, Gibbs Smith, 2010
  Very detailed ID and tasting notes on leafy weeds. Although US-based, most of it is relevant to Australia.

The Omnivore's Dilemma, Michael Pollan, Penguin, 2006
  Broad thoughts on the cultural and nutritional qualities of food, including wild foraged foods

Useful Weeds at Your Doorstep, Pat Collins, Total Health and Education Center, 1998
  For extensive medicinal uses and herbalists' recipes

Weeds of the South-East: an Identification Guide for Australia, F.J. & R.G. Richardson and R.C.H. Shepherd, 2011
  A seriously comprehensive aid to weed identification

WEBSITES

On the website of the Agricultural Research Service (US Dept of Agriculture), Dr Duke's Phytochemical and Ethnobotanical Databases are a source of detailed nutritional and pharmacological information about plants: www.ars-grin.gov/duke

Australia's Virtual Herbarium is a particularly useful website for discovering the distribution of weeds and other plants: http://avh.ala.org.au

Two US-based websites dedicated to 'Fighting Invasive Species, One Bite at a Time!' are: www.eattheinvaders.org and www.invasivore.org

The United States Dept of Agriculture's National Nutrient Database contains basic nutritional information for nearly 8000 food items, from microwave burritos to wild Alaskan sea cucumbers with several weeds in between: http://ndb.nal.usda.gov

Plants for a Future is an excellent database of over 7000 useful species of plants: www.pfaf.org

Weedy Connection is a useful weeds database and blog, also running Sydney-based weed walks: www.weedyconnection.com

The Intriguing World of Weeds on the Weed Science Society of America's website gives access to articles by the late US weeds history expert Larry Mitich, including cultural, edible and medicinal notes: www.wssa.net/Weeds/ID/WorldOfWeeds.htm

Herbal Medicine

Books

The Herbal of Dioscorides the Greek, Tess Anne Osbaldeston (trans), Ibidis Press, 2000
  See below for an online version of this influential two-thousand-year-old text.
Medical Herbalism: the Science and Practice of Herbal Medicine, David Hoffmann, Healing Arts Press, 2003
PDR for Herbal Medicines, Thomson Healthcare, Thomson Reuters, 2007
  Both these last two publications are impressively detailed herbal medicine references, suitable for practicing physicians and studious lay people alike.
A Modern Herbal, Margaret Grieve, 1931 (re-issued by Dover Publications in 1971) – See below for an online version of this classic of traditional plant usage.

Websites

The online version of *The Herbal of Dioscorides the Greek* is available in translation at: www.ibidispress.scriptmania.com/box_widget.html
The online version of *A Modern Herbal* can be found at: www.botanical.com

In the Garden

Books

The New Organic Gardener, Tim Marshall, ABC Books, 2012
Weed, Tim Marshall, ABC Books, 2011
  The second gives excellent advice on dealing with weeds in the garden, as does the first – in addition to providing good strategies for general organic gardening.

Weeds in Ecosystems and Regenerative Agriculture

Books

Back from the Brink, Peter Andrews, ABC Books, 2006
  Interesting strategies for land remediation, integrating use of weedy species
The New Nature, Tim Low, Penguin, Melbourne 2003
  A fresh look at the role of introduced species in Australian ecosystems
Weeds: in Defence of Nature's Most Unloved Plants, Richard Mabey, Harper Collins, 2011
  Fascinating ramblings on the history and psychology behind the very concept of weeds

WEBSITES

David Holmgren, co-originator of permaculture, is a deep and provocative thinker on topics such as weed ecologies, sustainability and global future scenarios, with writings available at: www.holmgren.com.au

Within the movement tackling invasive species, Professor Paul Downey from the University of Canberra has taken an impressively practical and scientific approach: www.canberra.edu.au/centres/iae/staff/downey

On these fascinating and challenging topics we've also been influenced by the thoughts of Peter Del Tredici, Mark A. Davis, Scott P. Carrol, John Dwyer and others. We have links to several of their articles on our website (www.eatthatweed.com).

OTHER LINKS

Very Edible Gardens is the Melbourne-based urban permaculture design business run by Adam and colleagues. It also offers an array of courses, veggie beds and chook systems: www.veryediblegardens.com

Permablitz is a collaborative, volunteer, backyard-makeover network (extending around Australia and the world) where you can get some hands-on experience, including dealing with weeds in the garden: www.permablitz.net

INDEX